Quantitative Vulnerability Assessment for Economic Systems

European University Studies

Europäische Hochschulschriften
Publications Universitaires Européennes

Series V
Economics and Management

Reihe V Série V
Volks- und Betriebswirtschaft
Sciences économiques, gestion d'entreprise

Vol./Bd. 3407

PETER LANG
Frankfurt am Main · Berlin · Bern · Bruxelles · New York · Oxford · Wien

Philipp Willroth

Quantitative Vulnerability Assessment for Economic Systems

Vulnerability and the Process of Recovery for Households and Companies in Phang-Nga and Phuket Provinces in Thailand

PETER LANG
Internationaler Verlag der Wissenschaften

Bibliographic Information published by the Deutsche Nationalbibliothek
The Deutsche Nationalbibliothek lists this publication in the Deutsche Nationalbibliografie; detailed bibliographic data is available in the internet at http://dnb.d-nb.de.

Zugl.: Hannover, Univ., Diss., 2011

D 89
ISSN 0531-7339
ISBN 978-3-631-63583-4

www.peterlang.de

Acknowledgements

This work is integrated in the project "Tsunami Risks, Vulnerability and Resilience in the Phang-Nga and Phuket Provinces, Thailand (TRAIT)" funded by the German Research Council (DFG). This project is supplemented by a project funded by the National Research Council of Thailand (NRCT) focusing on the development of risk management guidelines. Without the generous funding of the DFG and the close collaboration with the Thai partner project this work had never been possible.

Especially I like to thank my advisor Prof. Dr. Javier Revilla Diez for developing the project idea for the economic vulnerability assessment in TRAIT. He established an excellent network to the Thai project partners which allowed the comprehensive data collection this work can rely on. During the project he always gave conceptual support if necessary; but he also granted the space for developing new ideas and approaches. Furthermore I'd like to thank him for creating an excellent surrounding for writing this dissertation. Also I'd like to thank my colleagues at the Institute for Economic and Cultural Geography at the Leibniz University of Hannover for the uncounted discussions; especially I'd like to stress out Dr. Daniel Schiller who always has been open to discussions and shared his deep knowledge about Thailand with me. He also was the first to comment the draft of this thesis and gave several useful comments helping to finish this work. I also like to thank Priv.-Doz. Dr.-Ing. Jörn Birkmann for accepting the external examination of this thesis.

The comprehensive and holistic vulnerability assessment with the combination of remote sensing and survey data could only be possible with the strong collaboration within the TRAIT project. The coastal geography group in Kiel with Prof. Dr. Horst Sterr, Dr. Gunilla Kaiser and Hannes Römer focussed on the remote sensing aspects, the tsunami simulation and the ecological vulnerability assessment. I'd like to thank them for providing the remote sensing data, the tsunami simulation data and the technical support with the Geographical Information System. I also like to thank Prof. Dr. Ralf Ludwig from Department of Geography at Ludwig-Maximilans University Munich for his contribution to the automatic assessment of remote sensing pictures.

The comprehensive data collection could not been possible without the project partners from the Thai side under the project leader Dr. Somrudee Jitpraphai. Especially I'd like to thank Dr. Narumon Arunotai for organizing and conducting the surveys and the focus group discussions in the study area. With her excellent connections to the local authorities, population and businesses she made it possible to achieve such good results. Also I'd like to thank her research assistants who managed to interview the households and companies with the extensive questionnaire. Especially the commitment of Phasika Phalarak for organizing the survey in Phuket Province needs to be highlighted. The field mapping in the four study regions has been carried out by fifteen students from the Leibniz University Hanover under physical efforts in the tropical sun.

Table of contents

List of tables

List of figures

List of maps

1. Introduction

1.1 Background of research and research objectives

Safety is considered one of the most basic needs in human life according to Maslow's theory of human motivation (MASLOW 1970). This requirement must be fulfilled in order to develop further human activities and desires. A lack of information and knowledge, but also technological progress, might lead to decisions and strategies with risks that are often unforeseeable, but which are accepted in order to achieve individual or collective goals. Due to the advance of technological achievements and the possibility of externalising the outcomes of a risky strategy, or at least the risks unintentionally taken, more and more individuals, companies or nations are taking risks which might be classified in their appraisal as acceptable or tolerable. Therefore, Beck argues that we are living in a 'world risk society', where local hazards and risks span out to a worldwide scale due to the globalised interrelations in economic systems (BECK 2009:13). To deal with this uncertainty, risk assessment and management is becoming more and more important in scientific and public dispute (RENN 2008). Appraisal of risk depends on a variety of factors determined by human perception, livelihood characteristics and regional settings. Observing only the technical characteristics of natural hazards would not do justice to the complexity of the socio-economic sphere. Arising from research on poverty, access to entitlements and sustainability, the concept of vulnerability integrates the capacity of the economic sphere to cope with the consequences of such an impact, and places the impact in relation to the options and characteristics of managing and withstanding the impact. Mostly, vulnerability is defined as a combination of exposure to natural hazards or other shocks, sensitivity to being influenced by these shocks, and the capacity to cope with these external influences (BLAIKIE et al. 2007; ALWANG et al. 2001; CANNON et al. 2003; TURNER et al. 2003; IPCC 2001).

This study follows the definition of vulnerability proposed by TURNER et al. (2003:8074), which defines vulnerability as the degree to which a system or a system's component is influenced by an external shock. Here, vulnerability is defined from three perspectives, namely exposure, sensitivity and resilience, which determine the level of vulnerability. While exposure and sensitivity tend to increase the level of vulnerability, it is commonly argued that elements of the resilience perspective tend to lower or even compensate for the other elements. By taking the tsunami 2004 and its impact in southern Thailand as a reference event, this study investigates the influence of a singular and clearly defined external shock on a vulnerable society. Along the coastline of the Andaman Sea, the tsunami devastated wide areas of varying economic functions. Tourist areas as well as fishing and agriculture villages along the coastline of the Kingdom of Thailand were hit by the tsunami and destroyed. The majority of the population along the coastline was not aware of the risk of a tsunami in this area, since the last comparable tsunami in this area occurred between 550 and 700 years ago (JANKAEW et al. 2008:536). Five Years after the tsunami, most parts of the area are back to normal and can be seen as having fully recovered, which allows for a study of the vulnerable situation of the area prior to the tsunami and of its coping and recovery period afterwards.

TURNER et al. (2003:8077) name the application of the theoretical concept of vulnerability on statistical models in order to quantify the interrelations among the three perspectives in the economic sphere as one of the greatest challenges in this field. By using the unique and highly accurate data base of the project "Tsunami Risks, Vulnerability and Resilience in the Phang-Nga and Phuket Provinces, Thailand" (TRAIT), funded by the German Research Foundation (DFG), this study takes up this challenge. While several studies on vulnerability take the hazard as a static impact, this holistic study includes the physical impact set up by the characteristics of the tsunami scenario in combination with the determining settings within the units of study. This has been made possible by using a combination of highly accurate 15 cm remote sensing pictures and survey data which can be combined in a single model. Furthermore, the outcome of this model is verified in a triangulation process by results from focus group discussions. Against the background of the TRAIT project, this should enable the assessment of the vulnerability of the socio-economic sphere under the influence of different tsunami scenarios in order to derive risk management recommendations (cp. Figure 1).

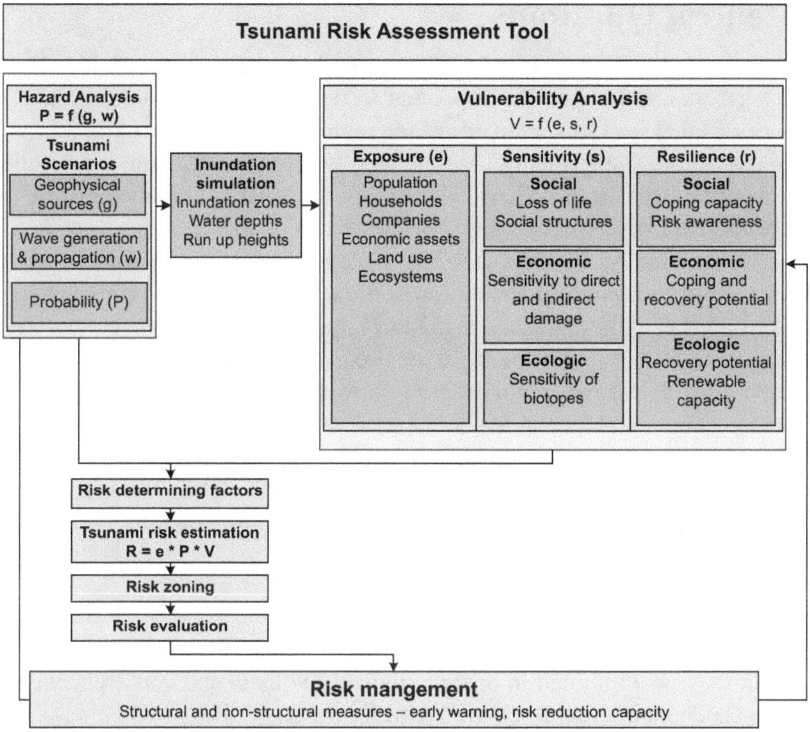

Figure 1: *TRAIT research framework (STERR et al. 2009:29)*

Along the coastline of southern Thailand, four communities were chosen as study areas for vulnerability assessment. These locations represent four different types of coastal communities typical for this area: Khao Lak represents a young developing tourism area; Ban Nam Khem is dominated by fishery and aquaculture; Thai Muang is an area with farming, incidental farming and a workforce related to the surrounding area of Khao Lak; and Patong Beach is a major centre of the Thai tourism industry with almost urban structures. These different settings provide the opportunity of comparing different levels of economic development and different capabilities of coping with natural hazards. While in the tourism centres, hotel complexes run by multinational companies are insured against the impacts of natural hazards, in the more rural areas, micro-scale businesses have only limited financial assets and almost no insurance. This great variety within the study areas should make it possible to develop generic and transferable methods which can be applied to other regions.

1.2 Central Questions

The aim of this thesis is to identify the vulnerability of the economic sphere of the Phang-Nga and Phuket Provinces and to reproduce the vulnerable state and the processes that lead to a recovery of the region's economic system. Aiming at the most accurate vulnerability assessment possible, the decision was made to model the processes on the micro-level, i.e. on the level of single companies and households. Due to shortages in secondary data on the chosen micro-level, a survey was conducted which was supported by remote sensing data in order to achieve a more comprehensive analysis of the region and the economic elements in it. As stated by TURNER et al. (2003:8077), one of the challenges in vulnerability research is to transfer the theoretical concept of vulnerability to an applicable and quantitative model. In order to do so, and to remodel the interrelations within the economic sphere in the study areas, a framework of guiding questions was developed supporting the major aim of this thesis in terms of developing a theoretical background, choosing the appropriate methods for analysis, defining a framework for the analysis and extracting policy implications based on the outcomes of the analysis:

Theory

T_1 What are the different streams in risk and vulnerability theories and how can they be combined in order to derive a holistic and comprehensive basis for economic vulnerability analysis?

T_2 Which perspectives of household or company activities and characteristics are considered in the models? How are these elements supposed to interact?

T_3 What are the effects of natural hazards, and especially tsunamis?

Methods

M_1 Which methods are suitable for reproducing the state and process of vulnerability?

M_2 In which way can remote sensing in combination with survey methods contribute to creating a database for exposure quantification on the micro-level and to supporting vulnerability analysis?

M_3 Which statistical method helps to transfer the theoretical constructs in a vulnerable system under the premise of identifying factors determining the vulnerability, and to reproduce their linkages?

M_4 Which indicators can be used to link micro-level data to building characteristics for identification by remote sensing?

M_5 Which indicators are suitable for representing the dimensions of vulnerability? Which proxy can be used to represent vulnerability?

Empirical Results

E_1 Which differences can be seen between the study areas in terms of directly and indirectly exposed assets?

E_2 What are the interdependencies between the elements of vulnerability? Which elements reduce vulnerability? Which increase the vulnerability of households? How are the relations quantified?

E_3 Which household and company groups can be identified as more vulnerable to natural hazards?

E_4 Which differences can be seen between the study sites?

Policy-guided

P_1 What kinds of adaptation measures have been taken on the personal level? Which measures have been taken by governmental and non-governmental organisations?

P_2 What implications do the findings of this work have for the endangered households and companies situated in potentially tsunami-prone areas? What are the special needs of the different groups?

1.3 Structure

The study is divided into six parts. The first part introduces the theoretical framework of this study derived from a review of current vulnerability literature (Chapters 2 and 3). In the second part, the data base and the methodological approaches for the assessment will be discussed (Chapters 4 and 5). The third part provides an introduction to the effects of the tsunami in 2004 on the macro-level and introduces the study areas (Chapter 6). In the fourth part, causal factors determining the vulnerability and their interrelations will be discussed and the results derived on the micro-level (Chapter 7 and 8). The experiences of the tsunami 2004 will also be applied to the current situation (Chapters 9). Finally, policy implications will be derived from the experience gathered in this study in relation to measures and strategies developed by the local population and agencies (Chapter 10).

The second chapter introduces the concept of vulnerability and how it can be adopted for the micro-level of households and companies in the study areas. This chapter discusses the advantages and disadvantages of this relatively new concept of vulnerability, which was introduced in the early 1970s (OEM 1972). Since this time, several approaches have been developed arising from different scientific or applied backgrounds, such as geography, economics, sociology and risk management, addressing the different topics and interests of these subjects. The results of these differing perspectives on vulnerability are included in a guiding framework for this study and answer the theory-related questions T_1 to T_3. This framework for the micro-level economic vulnerability assessment will be applied to the tsunami of 2004 as a reference event for vulnerability assessment in this study. The majority of these vulnerability concepts were developed as mainly theory-driven frameworks for qualitative research. Despite this, the major goal of this study is to adapt the concept for quantitative vulnerability assessment. In order to do so, there will be a discussion as to which indicators are suitable for representing the different perspectives of micro-level vulnerability considering the characteristics of households and companies in the study area. These indicators are to be derived from remote sensing data, secondary statistics where possible, or a household and company survey conducted in the study area. Additionally, these methods will be supported by remote sensing data in order to extend the coverage of the assessment to the entire group of households and companies in the region and to benefit from this method which is relatively efficient in terms of both time and cost. The following section will discuss which statistical method is suitable for reproducing and analysing the process of vulnerability in all its dimensions and quantifying the interrelations between the dimensions of vulnerability. This part addresses the methodological questions M_1 to M_5. After introducing the study areas and their exposure to and experience with the tsunami of 2004 on the meso-level in Chapter 6, the potential causal factors determining the vulnerability situation in this region will be discussed for each of the chosen study communities in Chapter 7. Based on the results of this chapter, a statistical vulnerability model for the micro-level will be developed in Chapter 8.

In Chapter 9, the model developed on a micro-scale basis for the households and companies surveyed will be applied to the entire community in its current situation, and the outcome will be compared with results from qualitative interviews in order to triangulate the quantitative results for verification. Finally, in Chapter 10, the empirical results will be summarised and a conclusion drawn in order to derive political implications for future natural hazards and to help to reduce the vulnerability of coastal communities.

2 Vulnerability of coastal regions to natural hazards: Theoretical framework

In all areas of the world, coastal regions are highly endangered by different kinds of natural hazards. Besides the natural exposure to natural hazards, various other factors, including the natural environment and anthropogenic factors, determine the degree of risk of these coastal areas. The holistic concept of vulnerability takes all these factors into account, combining the economic, social and ecological perspectives of a coastal region affected or endangered by natural hazards.

This is also a relatively new concept: vulnerability research is widespread across several scientific disciplines (RAYHAN 2008:6pp). In the following section, concepts from the fields of economics, geography, sociology and disaster management will be discussed and evaluated for their applicability to this study. In her study on core definitions of vulnerability, THYWISSEN (2006:449) found a 'babelonian confusion' due to different definitions arising from various scientific fields. The term vulnerability was first used in 1972 in a report to Congress by the Office of Emergency Preparedness of the Executive Office of the President of the United States. In this report, the term was used to mean the predisposition of people or communities and of business sectors to being affected by natural disasters. The report mainly stresses how important it is to identify the factors contributing to a disaster, and that these factors vary between the different kinds of disasters (OEM 1972). Following this, it was picked up by researchers from a wide range of fields such as economics, social science and development studies. This led to a vast number of terms used in vulnerability literature, including vulnerability, resilience, coping capacity, adaptation, risk, hazard etc. (BROOKS 2003:2). In general, there is consensus on accepting vulnerability as a starting point and basis for being at a certain level of risk when facing a hazard. Furthermore, there is agreement that vulnerability differs across space and between certain groups of individuals or business units. This system is not supposed to be static, but changes dynamically over time (VOGEL et al. 2004:3). In contrast to the field of damage potential analysis, covering physical damage of all kind, vulnerability is determined by characteristics of individuals or other study units which encompass social-economic indicators (CANNON et al. 2003). This already indicates that vulnerability is determined only partially by the hazard, and more by the characteristics of the affected region, household, company or individual. CANNON et al. (2003) named elements of households or individuals, such as physical well-being, livelihood circumstances (e.g. assets), income, qualification, social protection and networks, and institutions, which allow them to react to external shocks. This represents one of the first definitions of vulnerability, introduced in 1989 by CHAMBERS (1989:1), which defines vulnerability as the "exposure to contingencies and stresses and the difficulty which some communities experience while coping with such contingencies and stresses". In this definition, CHAMBERS pointed out that vulnerability consists of an external and an internal side. While the external side is related to the exposure to external shocks and stresses, the internal side represents the ability to defend and to cope with the external shocks and stresses. In general, social and economic vulnerability is determined by a lack of internal coping capabilities, for example low income, poor access to the health system, poor housing conditions and social networks. On the external side, institutions such as the government or informal institutions account for the ambient conditions leading to vulnerability and interaction with the internal side. This idea was picked up by BOHLE, who combined it with theories resulting from the field of poverty research (WATTS and BOHLE 1993; cp. Figure 2).

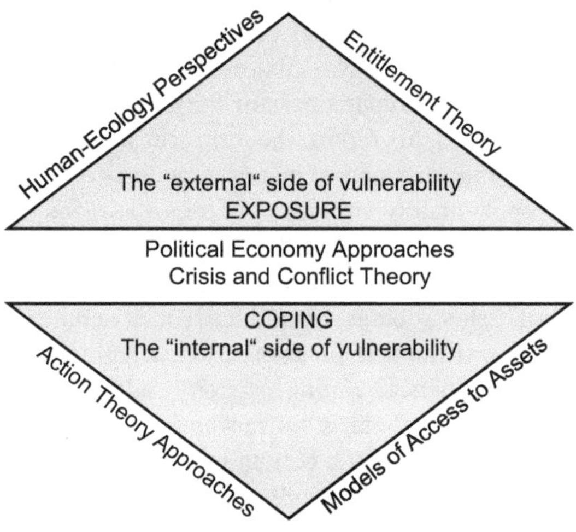

Figure 2: The two sides of vulnerability (Author's design, based on WATTS and BOHLE 1993:53)

In conclusion, it can be stated that vulnerability is a forward-looking concept which can be defined as the probability of experiencing a loss in the future in combination with a risky event. The degree of vulnerability depends on the characteristics of the risky event and the ability of the affected object to cope with the risk or mitigate its consequences (SIEGEL et al. 2003:363).

2.1 Approaches for economic vulnerability assessment

2.1.1 Vulnerability from the human perspective

The vulnerability assessment from the human perspective places its focus on the consequences for the individuals and households and the conditions and circumstances that influence their vulnerability in relation to their risk management, i.e. resilience and coping capacities. The entitlements approach has been strongly influenced by the capabilities literature introduced by SEN (1981), who looks at famine and hunger not only as the result of a lack of food ownership, but more as a result of a lack of entitlement to commodities. With his ideas, SEN shaped the focus of this vulnerability assessment approach by setting the key terms of entitlements, functioning and capabilities (GUIMARAES 2007:245). From this example, it follows that those individuals who are not entitled to basic functions lack the capacity to respond to a risk and must be described as vulnerable. This entitlement to basic functions takes into account the individuals' characteristics, their tangible and intangible endowments, and the social and institutional context that surrounds them. Tangible assets include land, labour, capital and savings, while intangible assets include social, institutional and political relationships. This set of characteristics leads to the capacity to respond to shocks and also to the individual exposure to external shocks. The key task is to identify the factors that explain the capacity of the household or individual to respond to shocks and how this interacts with different modalities of risks.

The vulnerability to entitlement approach can be split up into at least two sub-strands: the 'asset-based approach' and the 'sustainable livelihood approach'. The 'asset-based approach' focuses on the household's asset portfolio. It concentrates on the analysis of the stock of assets accumulated by the household. These include access to tangible assets as well as intangible assets. "Vulnerability is therefore closely linked to asset ownership. The more assets people have, the less vulnerable they are, and the greater the erosions of people's assets, the greater the level of insecurity" (MOSER 1998:3). In his study on urban economic vulnerability, Moser surveyed households about their accumulation of assets and their reaction to external shocks by using these assets to cope with the consequences. In detail, there was a focus on labour, human capital (i.e. health status, skills and education), productive assets (i.e. housing conditions), household relations (i.e. mechanisms for pooling income and sharing consumption) and social capital (i.e. reciprocity within communities and between households). The study used income to measure poverty and poverty dynamics by using longitudinal panel data. It could be observed that the households reacted to income declines by using different assets from their portfolio to cope with the new situation. In detail, it showed that the households mobilised additional workforces, i.e. women or children, depending on their educational level, to generate income to compensate for the decline. This also shows how important household ties are in their function as 'shock absorbers' (MOSER 1998). Despite this, the study does not provide a clear threshold for a minimum level of accumulated assets that allows the household to cope with the external shocks, and does not determine which special asset leads to a non-vulnerable livelihood.

The second strand of literature which can be traced back to the entitlement approach is found in the sustainable livelihood literature (cp. Figure 3). This concept was introduced by CHAMBERS (1989) in the 1980s. A sustainable livelihood comprises "the capabilities, assets (including both material and social resources) for a means of living. A livelihood is sustainable when it can cope with and recover from stresses and shocks and maintain or enhance its capabilities and assets both now and in the future, while not undermining the natural resource base" (SCOONES 1998:5).

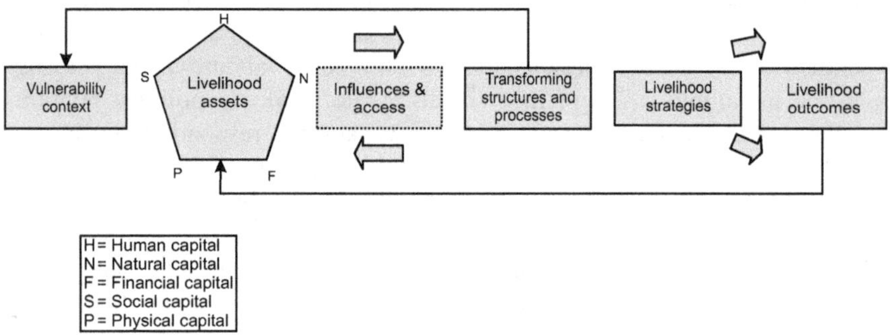

Figure 3: Sustainable Livelihood Framework (Author's design, based on SOLESBURY 2003:11)

The approach attempts to observe the well-being of the households from a holistic and dynamic perspective. Vulnerability, in this context, refers to the ability of people to avoid, to withstand or to recover from the harmful impact of a shock (BLAIKIE et al. 2007). It understands vulnerability as a result of three components, namely the exposure to shocks, the sensitivity of the system to the exposure, and the capacities to cope or respond, including the consequences and attendant risks of slow recovery (TURNER et al. 2003; GUIMARAES 2007). Livelihood resources influencing the vulnerability are deemed to consist of basic material and social, tangible and intangible assets that people have in their possession (SCOONES 1998:7pp):

- Natural capital: natural resource stocks (soil, water, air etc.) and environmental services from which useful resources and services can be derived
- Financial capital: capital base (cash, credit/debt, savings, remittances) and other assets (productive and non-productive assets)
- Human capital: skills, knowledge, health and other physical capabilities
- Social capital: social resources (networks, relations, affiliations) on which people can rely
- Physical capital: public infrastructure (transport, water supply, energy, communications) and private infrastructure (housing)

The household can reach a sustainable livelihood by varying this capital in different combinations; some capital can even be substituted (Scoones 1998). Depending on the combination of the household's assets, different livelihood strategies are identified. For example, intensification or extension of the household's income portfolio depending on the access and possession of financial assets and human capital. Another strategy would be the diversification of the household's livelihood. This means developing a wide income-earning portfolio to cover all types of shocks.

2.1.2 Operationalisation of vulnerability from the economics perspective

In economics literature, vulnerability is conceptualised as a process in which a household with a given set of conditions is affected by an external shock. Households will be defined as vulnerable if they are pushed under a poverty line as a result of the external shock. In this case, the terms risk and uncertainty are picked up by the economists' concept of 'vulnerability to poverty', which are also in use in the 'entitlement theory', the asset-based approach' and the 'livelihood theory'. As a consequence, vulnerability is a combination of the external shock and the household's characteristics (Siegel et al. 2003:355). The focus is placed on the identification of indicators concerning the household and its surroundings that determine the reduction of household income beneath the welfare threshold. Usually, metric indicators are used to measure the household characteristics in monetary terms or other metric indicators (Chaudhuri 2003; Mausch 2010). Poverty and vulnerability can be related in several ways, since poor people are more likely to be influenced in their livelihoods by external factors and, in addition, poverty can be an outcome or consequence of a vulnerable status.

2.1.2.1 Poverty Dynamics

A large amount of literature has been published on poverty dynamics focusing on the fact that poverty is not a fixed status, but that households can slide into poverty or pull themselves out of this state. In general, the authors argue that shocks affect individuals or households in different ways, and that their probability of falling under a certain welfare line is determined by the degree to which they are exposed to a risk and their capability to react (GUIMARAES 2007:237). CHAUDHURI (2003:4) defines vulnerability to poverty "as the ex-ante risk that a household will, if currently non-poor, fall below the poverty line, or if currently poor, will remain in poverty". This causality can be expressed in the formula $v_{ht}= Pr_{(ch,t+1 \leq z)}$, putting the probability of a household's consumption level at t+1 in relation to the poverty line z. The expression of the household's consumption in relation to the time t shows an important issue in this concept of poverty dynamics. Vulnerability is always understood as a forward-looking concept. This reveals the problematic nature of this approach. While poverty is a current status of an individual, it is assumed to predict the future status of this individual when he/she is affected by an external risk. The World Bank defines the 'vulnerability to poverty' concept "as the probability or risk today of being in poverty or of falling into deeper poverty in the future" (COUDOUEL et al. 2002:54). The aim of this research strand is to predict the level of probability of falling beneath the poverty line. As an example, one can consider two individuals: the first is employed as a public servant with a stable job and no probability of being fired; the second person is a day labourer without a contract. In time t, both individuals receive the same amount of income, and therefore have the same welfare status. Nevertheless, their vulnerability statuses differ considerably. With the appearance of an economic crisis, the second individual has a much higher risk of losing his/her employment and might fall beneath the welfare threshold.

The 'vulnerability to poverty' literature can be separated into two groups. The first group is more related to measuring the potential welfare loss in the future, i.e. the 'vulnerability as uncertain welfare', while the second strand of literature focuses more on the inability to respond to shocks due to a lack of individual, social, institutional or location assets, i.e. 'vulnerability as a lack of entitlements' (GUIMARAES 2007:239).

2.1.2.2 Vulnerability as uncertain welfare

The objective of 'vulnerability as uncertain welfare' is to predict today the probabilities of an individual's welfare for tomorrow, taking into account its characteristics and the probable risks that may occur. Individuals or households are defined as vulnerable if they are pushed below the welfare threshold by an external shock due to the fact that they are not able to cope with this shock because of their specific characteristics. This vulnerability measurement is expressed by the probability of falling below the welfare threshold. In this approach, risk is considered to be endogenous to the welfare equation due to the assumption that future welfare results from the characteristics of the individual or household and the expected outcomes of this set of characteristics.

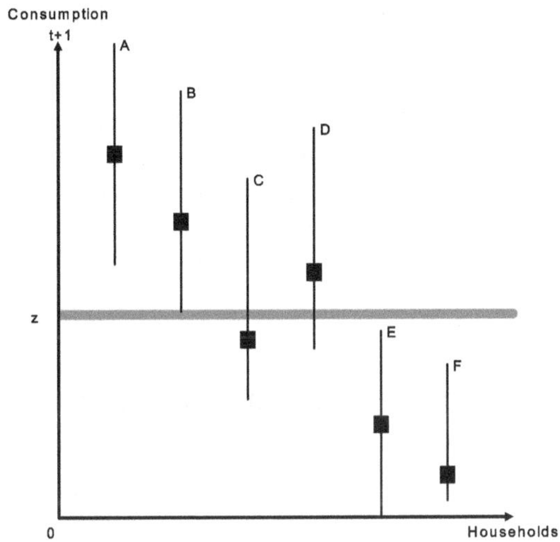

Figure 4: Expected level of consumption (Author's design, based on GUIMARAES 2007:242)

The approach follows the theories of microeconomics assuming that the individuals aim to maximise their utility and that their expenditure reflects the marginal value of the people's interest. These assumptions are transferred in mathematical models representing the individual's behaviour. This approach follows the assumption of probability and implies that risk events and their outcomes follow an uncertainty pattern. Following a non-deterministic setting, CHAUDHURI et al. (2002) proposed a model taking two periods into account, defining vulnerability as: $V_i = \int_0^z p(x,z)(dF(x))$. In this model, the poverty line is represented by z, x is the consumption and F(x) is the distribution of consumption at time t that is taken as given, and also captures the exposure of the individual and his/her response to risks. The poverty measurement is represented by the equation $p(x,z) = \left[\frac{\max(z-x)}{z}\right]^\alpha$. The variation of household expenditure is based on assumptions and statistical assumptions regarding the future welfare position. As an outcome, the expected consumption of each household is calculated and possible changes resulting in different risks can be estimated, showing the probability of each household of falling beneath the poverty line (cp. Figure 4). The square represents the household's expected consumption in t+1, while the vertical line represents possible states of consumption in t+1. For example, household C expects consumption beneath the poverty line, although consumption above this poverty line is still possible depending on the state of the surroundings. The households E and F in this example will definitely fall below the poverty line in t+1 and can be called vulnerable, while household D has only a marginal chance of falling beneath the poverty line, and is therefore only slightly vulnerable.

ADGER (2006:272) criticises this approach for disregarding the physical and ecological dynamics of risk. The focus is placed solely on the consumption or welfare of households as an outcome or precondition of vulnerability. One positive feature in terms of achieving a comprehensive analysis of vulnerability is the fact that the approach integrates the sustainable livelihood idea, which comprises the capabilities, assets and activities that can lead to the well-being of a household.

2.1.3 Risk analysis or natural hazards approach as an element of vulnerability

This important element of vulnerability arises more from the technical or civil engineering approach and defines risk as an outcome of occurrence probability and the amount of damage (MEYER 2005:28). Here, risk is defined for a specific event, with its specific characteristics describing its severity. For example, a storm flood can be described by the height of the tide (Thw), the sea disturbance and the specific probability (p_i) of this event occurring and causing a specific amount of damage (D_i). Therefore, the risk is defined as $R_i = p_i*C_i$. Probability of occurrence and damage ratio vary between different types of events. Applying this formula for event scenarios, the possible outcomes of risk can be calculated. According to the definition of risk, two steps have to be taken for a full risk analysis (REESE 2003:24pp). In the first step, the threat of the hazard has to be analysed, resulting from the probability of the occurrence of the event and the failure of the counteracting protection system. Building upon this probability analysis, an ex-ante damage analysis is conducted in order to evaluate the potential damage of this specific event. For the first step, the specific event has to be defined. This can be either a scenario which is likely to happen or a reference event that has already occurred. The characteristics of this event and the topographic characteristics of the area are the main influences for the damage analysis (MEYER 2005:31). In the second step, the damage potential or the actual amount of value for the areas possibly affected needs to be evaluated. Combining these two elements, the characteristics of the event, counteracting protection systems and the damage potential can be calculated. The actual damage that occurs can be expressed as a mathematical function of damage ratio and damage potential. The most important influencing factor on the damage ratio in the case of flood events is the inundation height (REESE 2003:199). Of less importance is the duration of the flooding in terms of direct damage (BEYENE 1992:13). Another influencing factor is the velocity of the water (BWK 2001). In general, not all of the influencing factors could be considered for the analysis. Most studies only consider the most important influencing factor, namely the inundation depth (SMITH and WARD 1998:41; BEYENE 1992:14). To assess the overall damage, two approaches are possible. MEYER and MAI (2003) calculated the amount of damage as a percentage of the assets. BEYENE (1992), on the other hand, combined functions of inundation heights and damage using absolute amounts of damage.

The primary goal of this approach is to predict the damage arising from natural hazards to the livelihood of households, company property or government property. Fields of use for this method include cost-benefit analysis for coastal protection measures, insurance issues, and studies in the field of climate change measuring the consequences of the outcome of climate change (IPCC 2007:666pp). In terms of vulnerability analysis, this approach only focuses on the monetary outcomes of natural hazards, including components of exposure to natural hazards, external sensitivity (namely coastal defence and early warning systems) and internal sensitivity in the case of the impact of the hazard determined by its characteristics, such as inundation height and others. Neither the resilience perspective nor internal sensitivity are considered by this approach, but this is not its focus. Nevertheless, the integration of damage analysis into vulnerability analysis could offer the possibility of comparing the interrelations of resilience and sensitivity with different levels of physical damage. A closer examination of vulnerability analysis was undertaken by BURTON et al. (1993). They summarised research on flood management, geo-hazards and man-made hazards, deriving lessons on individual risk perception through to international collective actions. One outcome of their synthesis was the acknowledgement that impacts differ between various groups within society. In particular, they found that for many natural hazards, the vulnerability of the human population is determined by where they reside, how they use natural resources and which resources they have for coping. This study already contains elements of TURNER's et al. (2003) coupled human environment approach and the more technical natural hazard idea of impact research.

2.1.4 Synthesis of physical and social vulnerability

The pressure and relief model developed by BLAIKIE et al. (2007) attempts to bridge the disparities between physical and social vulnerability by integrating social circumstances and exposure to natural shocks (Figure 5). According to the model, a disaster occurs when two opposing forces, i.e. a hazard and a vulnerable society, household or individual, confront each other, which can be expressed as Risk = Hazard x Vulnerability. They identify vulnerability as a chain of root causes, dynamic pressures and unsafe conditions (BLAIKIE et al. 2007). Root causes reflect the access and distribution of power in a society. These causes are related to economic, demographic and political processes within a society. These are channelled in dynamic pressures creating unsafe conditions, such as population growth, rapid urbanisation, deforestation, or a lack of training and appropriate skills among the population, as well as deficiencies in local markets or investments. Root causes and dynamic pressures manifest themselves in unsafe conditions, forming the vulnerability of a society, household or individual. BLAIKIE et al. (2007) refer to several examples such as dangerous settlement locations, low income level, lack of local institutions or the absence of disaster preparedness (cp. Figure 5). The most important advantage of the 'Pressure and Release model' is that it attempts to explain the development of unsafe and vulnerable conditions (DE VILLAGRÁN LEÓN 2006:18).

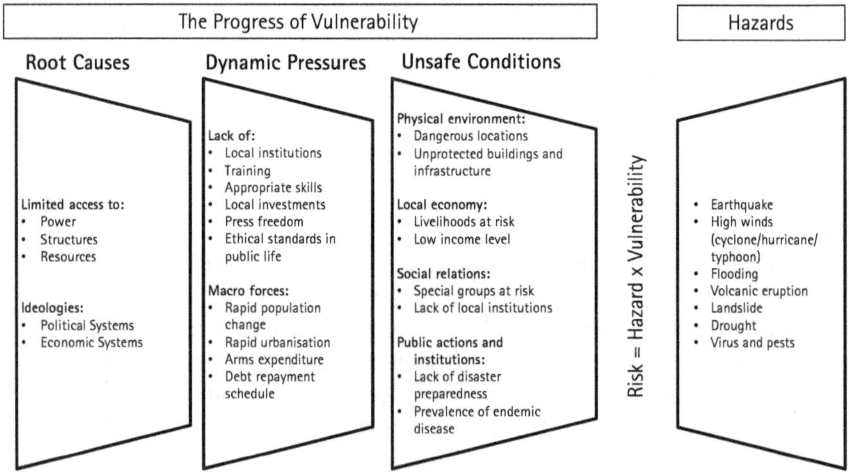

Figure 5: Pressure and release model (Author's design, based on BLAIKIE et al. 2007:47)

While giving equal weight to the social and physical part, the model is insufficient in providing a systemic view of the mechanisms and processes of vulnerability (ADGER 2006:272).

2.2 Integrative frameworks

Building upon the ideas of entitlement theory, damage analysis and the 'pressure and release approach', the system-oriented frameworks of vulnerability attempt to understand vulnerability in a holistic way, covering natural, social and economic systems together (cp. Figure 6).

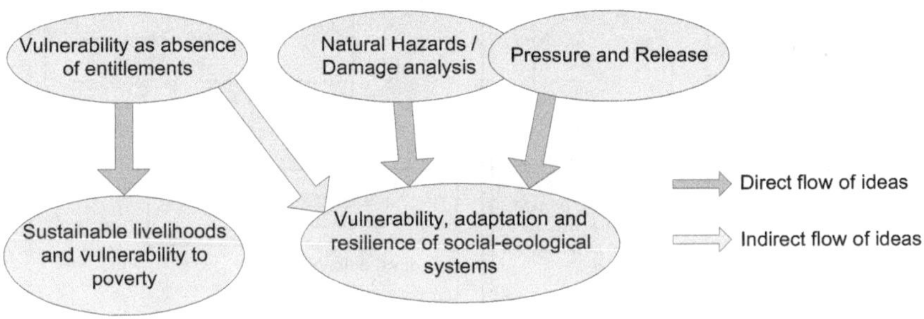

Figure 6: Evolution of vulnerability frameworks (Author's design, based on ADGER 2006:271)

In general, these attempts can be characterised as encompassing the perspectives of exposure, sensitivity and coping capacity "Vulnerability is often counter posed with capacity to anticipate, cope with, resist and recover from hazard impacts." (WHITE et al. 2004:15; TURNER et al. 2003; CLARK et al. 2000; ADGER et al. 2004). Exposure in general can be understood as elements in a certain area that are at risk when a specific stress factor impacts upon this area (UNDP-BCPR 2004). The second dimension of vulnerability, sensitivity, defines the degree to which an exposed element is affected by the impact of a natural disaster or external shock (CLARK et al. 2000). The final dimension of vulnerability is the coping capacity – the ability of an exposed element to absorb or resist the impact that could result from a disaster by using available resources and capabilities (UN-ISDR 2004; ADGER et al. 2004).

TURNER et al. (2003) follow a broader assessment of vulnerability in their framework and attempt to grasp the totality of coupled human environment systems embedded in their surroundings on the meso and macro-levels. This framework seeks to analyse the elements of vulnerability, i.e. exposure, sensitivity and resilience, and to assess the linkages between them (cp. Figure 7).

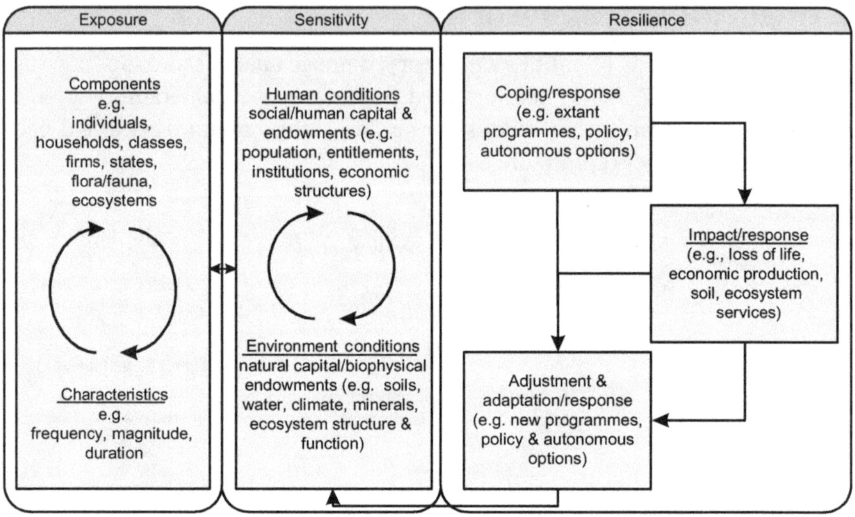

Figure 7: Vulnerability framework and its components according to TURNER et al. (Author's design, based on TURNER et al. 2003:8077)

In each of these elements, the linkages between human and environmental conditions play an important role. The element of exposure is shaped by the conditions of individuals, households or firms and the characteristics of the interfering shock, for example the frequency, magnitude or duration of an earthquake. In contrast to the other vulnerability frameworks introduced above, shocks are not only seen as an exogenous influence, but also as a consequence of the circumstances in which the household or individual is living. The shock can derive from these linkages, but is also influenced by outside conditions. The element of sensitivity, consisting of the human side of capital and endowments, i.e. financial and human capital, is linked to natural capital on the environmental side. This shows again the closed linkage to the sustainable livelihood framework and its basis, the entitlement approach. Exposure and sensitivity are not separated, but also interact and influence each other. Also, the element of resilience is linked to the other parts or elements of vulnerability. The human conditions as well as the exposed components influence the capability of the household or individual to cope with or respond to a shock. This coping mechanism influences both the outcome of the shock and the adaptation process following the shock. TURNER et al. (2003) make particular mention of the embeddedness in processes on the meso and macro-scales. The processes have an influence on the human or environmental conditions caused by political decisions as well as coping or adaptation mechanisms driven by governmental or non-governmental agencies. Like the asset-based vulnerability approach or the sustainable livelihood literature, the framework by TURNER et al. (2003) does not provide clear measures and indicators for identifying vulnerable individuals or households. Its main purpose is to illustrate "the complexity and interactions involved in vulnerability analysis, drawing attention to the array of factors and linkages that potentially affect the vulnerability of the coupled human-environment system" (TURNER et al. 2003:8077). The conceptual BBC-Framework developed by Birkmann and Bogardi and Cardona goes one step further, underlining the dynamics of vulnerability (DE VILLAGRÁN LEÓN 2006:17). They stress that vulnerability assessment has to take into account the feedback loop system by simultaneously looking at exposed and sensitive elements, coping capacity, and potential intervention tools (cp. Figure 8).

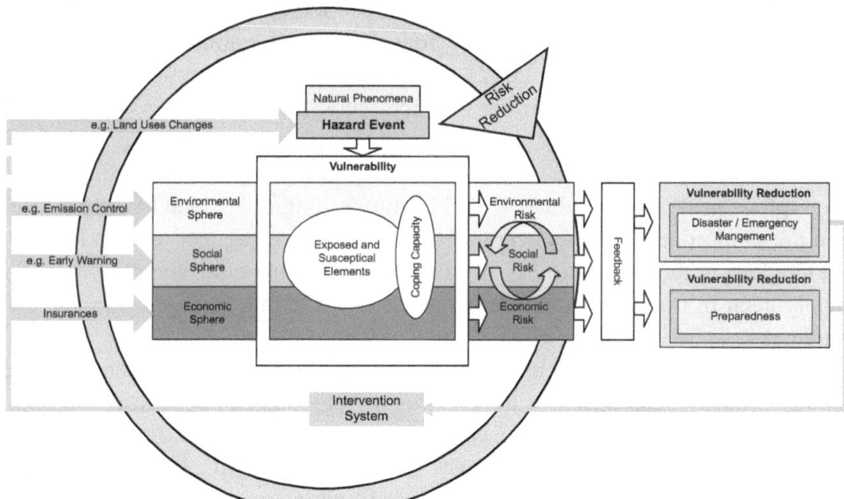

Figure 8: BBC Model for vulnerability assessment (Author's design, based on CARDONA 1999; BOGARDI and BIRKMANN 2004; CARDONA 2001)

As the framework is based on the concept of sustainability, it takes the economic, social and environmental sphere into consideration and addresses the linkages between these spheres. In contrast to the vulnerability to poverty concepts, the sustainable livelihood or the pressure and release model, this framework acknowledges that the environment itself can be vulnerable to external shocks, for example human-induced shocks or natural hazards (BIRKMANN and FERNANDO 2008:86). Moreover, it emphasises that vulnerability assessment cannot be reduced to observing the disaster, its outcome and the disaster management process, but requires acknowledgement of the lessons learnt from this process, resulting in mitigation strategies and strongly influencing the vulnerability afterwards.

2.3 Vulnerability assessment for various-scale regions

The majority of studies aiming at vulnerability analysis focus on the household level, with the 'vulnerability to poverty' concept being particularly strongly linked to the micro-scale level of household vulnerability (LIGNON and SCHECHTER 2003; GUIMARAES 2007). However, a great variety of studies exist which focus on vulnerability on the country level, mainly concerning small island states (TURVEY 2007; LIOU and DING 2004), as well as the hotspot analysis conducted by the World Bank assessing vulnerability on a scale defined by a cartographic 'grid' delimited by 1° x 1° latitude/longitude (DILLEY 2007). Between these approaches, the concept of 'vulnerability of regions' can be located. It acknowledges that vulnerability of households can have different sources. As typically seen in previous literature, risks are divided into idiosyncratic and covariate risks.

The vulnerability of regions concept follows the idea that vulnerability on the household level is strongly dependent on regional characteristics that can lead to a vulnerable situation in the case of covariate shocks (NAUDÉ et al. 2009; BIRD et al. 2007; WEBBER and ROSSOUW 2010). The objective of studies aiming to assess vulnerability of place is to understand why some places are more vulnerable than others. In this study, place is closely related to geographic dimensions such as location, size of area and physical environment of the place, and the social and economic interactions occurring in this region. TURVEY (2007) divides vulnerability of place into three elements: the social element, concerning people or communities located in areas at risk, the spatial element or vulnerable places, and the temporal element expressing the varying vulnerability at different time configurations. Following this argumentation, NAUDÉ (2008) refers to the fact that regional factors can have a direct impact on households' incomes and well-being capabilities, and therefore influence the vulnerability of households by their specific configuration of factors. Overall, the place vulnerability can be summarised as a function of socio-political determinants and economic geography in a given geographical region.

TURVEY (2007) introduced three causal factor groups, namely economic, geographical and socio-political factors explaining the specific situation of place vulnerability:

Economic factors

- Vulnerability to external economic shocks
- Fragility due to intrinsic factors
- Susceptibility of domestic economy to extreme events
- Vulnerability to routine risks in 'everyday life'

Geographic factors

- Vulnerability to natural disasters
- Vulnerability from location disadvantages – peripheries, ghettos, slums
- Endangered zones and impaired habitats – typhoon and hurricane belts, polluted areas
- Vulnerability to structural weakness – dependence, food insecurity, powerlessness, poverty and lack of response capabilities
- Fragility of ecosystems and physical environment
- Rural, urban, sectoral and communal space of vulnerability

Socio-political factors

- Enforced vulnerability of population – forced labour, forced resettlement and uprooting, economic sanctions, ethnic cleansing and regions of misrule
- Vulnerability to interpersonal forces – population factors, economic, cultural and environmental conflicts

This catalogue of factor indicators provided the basis for a sub-regional study on vulnerability. By averaging the indicator values a vulnerability index has been derived. This index points out that besides internal characteristics of households or companies, meso or macro-level characteristics such as "susceptibility of domestic economy" also influence the livelihoods and business activities.

2.4 Risk management

As argued previously, vulnerability can be reduced by increasing the counterbalance (the resilience as the capacity to deal with a shock), by decreasing the impact of the shock, or by avoiding the shock. In the relevant literature, three kinds of risk management possibilities are discussed for reducing the vulnerability which are related to the time at which they are performed in relation to the shock, the level of formality or the player (HOLZMANN and JØRGENSEN 2001).

The category related to the time after which the risk measures are implemented is oriented towards the occurrence of external shocks (cp. Figure 9). Before a disaster occurs, prevention strategies can help to reduce the probability or the consequences of the impact of an external risk. In the case of poverty-related risks, this can include increasing people's income or reducing income variability through various strategies. For preventing poverty risks, strategies can be followed which provide education and training, public health, environmental protection measures or other strategies increasing the people's and institutions' abilities to prevent potential external shocks. In the case of natural hazards, i.e. tsunami, storms etc., risk reduction can be achieved by building physical flood barriers, relocating the population or carrying out other protection measures. These ex-ante measures all aim to reduce the risk or decrease the effects of the potential external shock.

From the ex-post perspective, resilience measures and strategies help to deal with the outcome of risks. These strategies for coping and mitigating the consequences of the shock that has just occurred include strategies implemented at the household or company level, such as borrowing, migration or the reliance on public or private transfer payments. These measures strongly depend on a household's capabilities and skills that are available and can be used for this purpose. While coping measures are targeted towards reducing the impact of a shock that has just occurred, adaptation strategies, which are also an outcome of the shock, are more targeted at future events (BIRKMANN 2011:5). In contrast to prevention strategies, which attempt to reduce the probability of the occurrence of a shock, adaptation strategies reduce the potential outcome of the disaster if it were to occur in the future. Adaptation strategies encompass several forms of strategies for extending the household or company resilience capacities, such as income or sales diversification, informal or formal insurance mechanisms, or complete relocation of settlements.

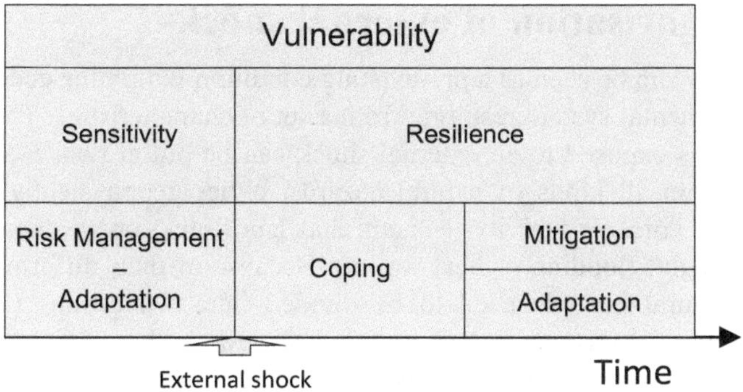

Figure 9: *Time dimension of coping and adaptation (Author's design, based on* SCHNEIDERBAUER *2008:17)*

In addition to the distinction concerning the time frame of the intervening strategy, the level of formality also divides the risk management strategies. In societies lacking market institutions and public provisions, informal arrangements for risk management are often widespread. These arrangements include informal borrowing, buying real assets as a substitute for saving, or storing goods for future consumption. These measures are often characterised as being less effective and tend to be more expensive in terms of opportunity costs (MORDUCH 1999). Market-based arrangements, such as the financial system providing bank deposits, lending and insurance contracts, are more effective. Also publicly mandated arrangements are possible such as the government providing risk transfer in the form of insurance programmes, for example unemployment benefits, old-age pensions etc., or social assistance, for example cash or in-kind transfers. This kind of risk management is highly dependent on the available fiscal resources and the administrative capacities available in the government.

The individual player's role also needs to be considered in risk management. Players in risk management can vary from individuals and households on the lowest scale to international organisations on the highest scale, such as the United Nations, the World Bank or the International Monetary Fund, providing risk management. The level of information varies in an asymmetric way along with the scale of players. The affected household or individual holds all the essential information about the ongoing risk. At the other end of the scale, the international institutions lack information about the risk for the individual household or person (HOOGEVEEN 2000).

2.5 Categorisation of external shocks

Vulnerability can be seen as a pre-existing condition within the economic, social and environmental system resulting from a set of characteristics. This vulnerable system that is exposed to an external shock can be put at risk. External shocks can arise from all kinds of natural hazards, either geophysically related (e.g. earthquakes, volcanic activity, tsunami and landslides) or weather-related (e.g. storms, drought, flooding or heat waves). Because of their differing impacts on systems, natural hazards need to be divided into two groups (WHITE et al. 2004:13). The first group contains natural hazards with a rapid onset. This includes earthquakes, volcanoes, fires, landslides, tsunamis and flash floods. All of these hazards only allow a very limited time to react and have almost no opportunity for warnings. On the other hand, there are 'slow onset' hazards that develop over a longer time period, such as droughts or rising sea level. Here, an early warning is much easier and external help can be organised more carefully. External economic risks such as price changes, or internal risks such as health problems of family members, death, crop loss or other events, can also be included as external shocks (LÜBBEN 2009:19). Another characteristic dividing risks is the scale of people affected by an external shock. HEITZMANN et al. (2002:24) divide risky events according to the "degree of its correlation among individuals, households, communities and regions". Shocks affecting households, for example unemployment or death of a household member, can be uncorrelated among individuals and can only affect specific individuals, while others affect a group of households or entire communities (e.g. earthquakes, economic crisis). According to HEITZMANN et al. (2002:24pp), economic risks such as business failure, or health risks such as illness, can be identified as 'idiosyncratic events', while other risks, especially natural hazards, are more 'covariant events' (cp. Table 1).

Table 1: Categorisation of external shocks (Author's own design)

	Idiosyncratic shocks		**Covariant shocks**	
	Rapid-onset	**Slow-onset**	**Rapid-onset**	**Slow-onset**
Natural			Earthquakes, Floods, Tsunami	Drought, Sea level rise
Economic	Unemployment	Business failure	Harvest failure, Economic crisis	
Health	Illness, Accident, Disability		Epidemic	
Life cycle	Birth, Family break-up	Old-age, Death		
Social	Crime, Domestic violence		Terrorism, War	
Political	Discrimination			Coup d'état
Environmental			Nuclear disaster	Pollution

When this external shock is combined with a system in a vulnerable condition, the shock can become a disaster for the affected system (DE VILLAGRÁN LEÓN 2006:9). This can be expressed in a mathematical formula where the combination of hazard and vulnerability leads to Risk (Risk = Hazard x Vulnerability) (UN-ISDR 2004).

3 Vulnerability framework used in this study

As seen before the major approaches of vulnerability research can be divided into two strands (cp. Figure 6). On the one hand, the concept of 'vulnerability to poverty' uses an income threshold to define a vulnerable state. If the household is likely to fall under this income threshold, the household can be called vulnerable. On the other hand, there are the frameworks resulting from the socio-economic research fields which are more focused on describing the interactions of exposure, sensitivity and resilience in a more qualitative and holistic way.

The aim of this study is to create a quantitative measurement tool for vulnerability which is suited to the special situation of the 2004 tsunami and integrates the elements of vulnerability discussed above. The concept of vulnerability to poverty is not applicable for the special situation in southern Thailand after the tsunami due to the enormous efforts undertaken by governmental and non-governmental agencies to support local households and businesses in their recovery efforts. As a consequence, the poverty rate in this relatively high income area of Thailand is still significantly lower than the average of the whole kingdom. The main characteristics of the 2004 tsunami in southern Thailand result from its special nature as a rapidly occurring and covariant risk affecting all households to the same degree, with differences depending only on their location relative to the coastline and their exposed assets. Due to the fact that tsunamis were not known in this area before, no special risk management strategies on the level of households and/or companies or on the meso-level had been developed before the event occurred. By using the concept of vulnerability proposed by TURNER et al. (2003) and further developed by BIRKMANN (2006), the three dimensions determining the vulnerability will be assessed. This concept will be used for this study to assess vulnerability and recovery from the effects of the 2004 tsunami. In the following section, the individual elements and the indicators forming these elements will be discussed. Furthermore, the more technical approach of damage analysis, or the natural hazards perspective, will be integrated into the conceptual framework of this study.

3.1 Elements of economic vulnerability

SCHÄTZL (2003:99pp) defined the components determining the economic system of regions in a very comprehensive way. According to the macroeconomics literature, the economic system of a region relies on supply and demand. Supply is determined by the production potential, which depends on the quality and quantity of the production potential in the region, i.e. labour, assets and land. The demand side depends on the private and governmental demand for consumption and investment goods. Both sides, supply and demand, are influenced by the technological progress achieved in the region, the spatial pattern, the pattern of the business sector, infrastructure, and the political and social system. Here, the spatial pattern represents the given distribution of economic activities within the region. Additionally, the economic performance in a region depends on the interaction with other regions. Exchange and trade of mobile production factors, i.e. labour and capital, and of goods and services influences the economic activity and performance of a region (cp. Figure 10).

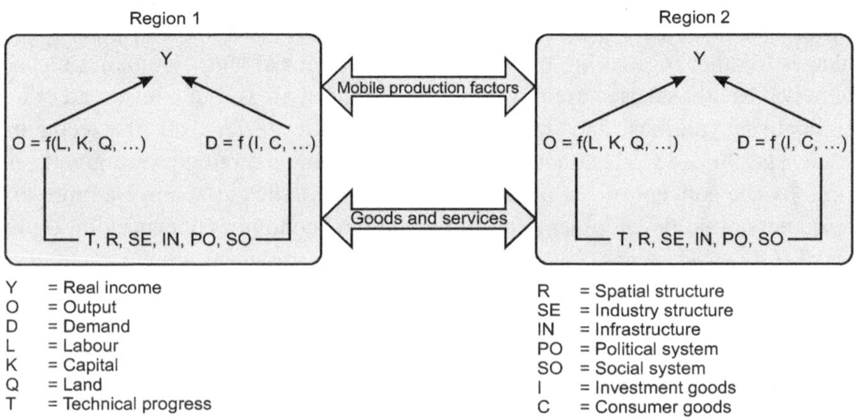

Figure 10: Elements of an economic system (Author's design, based on SCHÄTZL *2003:103)*

This model by SCHÄTZL has been developed to explain the reasons for economic growth, but it is also highly suitable as a framework for the determination of the factors leading to economic vulnerability. In combination with the sustainable livelihood framework and the vulnerability framework by TURNER et al. (2003), a comprehensive vulnerability model can be created. The exposure towards external shocks measures the "degree to which a human group or ecosystem comes into contact with particular stress" (CLARK et al. 2000:2). On the exposed side of the vulnerable economic system, all assets available in the region can be found. Here, SCHÄTZL mentions investment goods on the company side and consumer goods on the level of households which might be at risk when a natural hazard eventuates. The degree to which the exposed elements are affected depends on their sensitivity (cp. chapter 2.1.3). The special characteristics of a region include its industry structure, the spatial structure, the political system, the social system, the infrastructure and the technical progress available in the region. All these items have to be evaluated concerning the quantity in which they are available in the region, but also in terms of their quality. Finally, there is the resilience - the "ability of a system to absorb perturbations or stress without changes in its fundamental structure or function" (KASPERSON et al. 2005:253). Here, all characteristics of the system that reduce the effect of the external shock come into play. The sustainable livelihood framework in particular attempts to define these elements by referring to the different capital that can be used to cope with the effects of an external shock. Matching the elements mentioned by SCHÄTZL, the framework quotes the natural, financial, human, social, and physical capital which can help to reduce the impact.

Nevertheless, this system is not closed and focused on a limited area. Due to connections between companies or companies and consumers, interregional linkages to other regions are established. Connections among companies respectively wages and salaries are not limited to a closed areas, but can spread out to neighbouring regions and may have a multiplier effect on the economy (SCHÄTZL 2003:183). Thailand's regional development strategies are greatly inspired by the concept of regional economic growth poles to achieve a more dispersed economic development and to reduce the economic concentration on the capital (GLASSMANN and SNEDDON 2003). In order to decentralise the country's economy and to develop secondary economic growth centres in all regions of the country, five year plans were published by the National Economic and Social Development Board. In the third National Economic and Social Development Plan (1972-1976), nine regional economic growth poles in medium-sized towns in all regions of the country were designated. The implementation began in the fourth plan starting from 1977 (PARNWELL 1992:55). The idea behind this regional growth pole concept is that industrially formed agglomerations spread out growth impulses to the surrounding areas (SCHÄTZL 2003:183). From these sub centres, intraregional decentralisation processes are expected to spread out to the surrounding smaller communities until a stable regional system is achieved (RICHARDSON 1980). Therefore, inter-industry and inter-regional economic connections also need to be taken into account in an economic vulnerability model.

3.2 Conceptual framework for this study

This study builds upon the definition of vulnerability proposed by TURNER et al. (2003) which defines vulnerability as the degree to which a system or a system component is influenced by external stress, for example by a natural disaster or any other external shock. The theoretical construct of vulnerability consists of three interconnected dimensions. These three dimensions are exposure, sensitivity and coping capacity / resilience capability. Based on the theoretical discussion above, and especially on the basis of TURNER'S et al. framework and the BBC Framework, the framework developed for this study combines these three elements into one closed system and emphasises that vulnerability assessment does not stop at the assessment of disasters in the past (BIRKMANN and FERNANDO 2008:85). In fact, it is crucial to assess the process of vulnerability and to include the efforts made to reduce the vulnerability after the disaster by influencing vulnerability, coping capacity or interventional systems such as risk management or early warning systems (cp. Chapter 2.4). The conceptual framework relates the elements of SCHÄTZL'S model of an economic system to Turner's vulnerability model. The elements of an economic system were assigned to the three perspectives of vulnerability according to the concepts of sustainable livelihoods and the entitlement theory, as well as the conceptualisation of exposed economic assets by SMITH and WARD (1998). As exposure, all potentially directly and indirectly affected tangible goods and monetary flows in households and companies will be observed (cp. Chapter 3.2.1; cp. Figure 11).

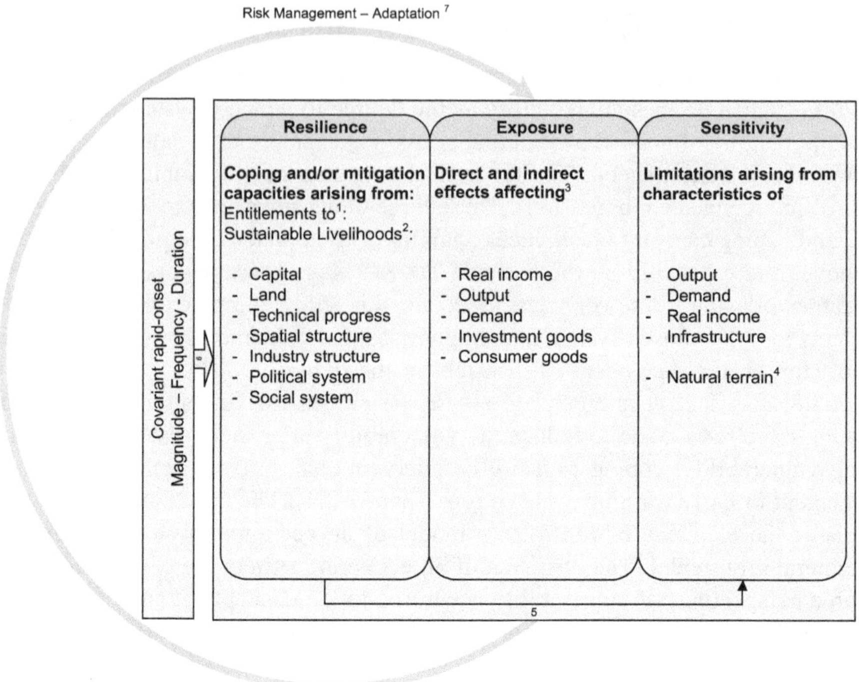

Figure 11: Conceptual framework (Provided by author, based on [1]SEN 1981; [2]SOLESBURY 2003; [3]SMITH AND WARD 1998; [4]Chapter 2.1.3; [5]TURNER et al. 2003; [6]BLAIKIE et al. 2007; [7]HOLZMANN and JØRGENSEN 2001; Elements of an economic system based on SCHÄTZL 2003)

The degree to which the exposed elements are affected depends on the related sensitivity-determining elements, namely the specific characteristics of the monetary flows or, in the case of a tsunami, the location of the geographically fixed assets. The final perspective of vulnerability, the resilience or coping capacity, consists of factors which result from entitlements to tangible or intangible assets for households and companies, or which are recognised as a basis for a sustainable livelihood. In the following section, these elements will be operationalised by suitable indicators in order to quantify the vulnerability on the household and company level. The quality and quantity of these elements will hereby play an important role. Governmental activities regarding the demand side are not included in the focus of this study and will not be evaluated in the subsequent work.

3.2.1 Exposure of households and companies

Exposure in general can be understood as elements in a certain area that are at risk when a specific stress factor impacts upon this area (UNDP-BCPR 2004). In the case of assessing the economic vulnerability, this refers to the value of economic assets that are at risk in this area or that must endure the certainty of loss (ESPON 2003; MUNICHRE 2002). Besides the households' or companies' property, monetary flows such as income or sales and demand are also at risk when a disaster occurs (cp. Figure 12). The majority of studies dealing with the impact of natural hazards are focused on the direct impact on property which occurs immediately after the event and results from the physical impact on property. The indirect effects are more difficult to assess, but no less important. These costs mostly emerge in the days, weeks or months following the event and are caused by the direct effects. Indirect losses occur mostly on a long-term scale, leading to loss of income or reduced economic performance of the region. Indirect costs include losses due to business interruptions, for example as the consequence of the interruption of infrastructure or if employees are unavailable for their own personal reasons. Other indirect costs result from the reduction of tourism revenues (HEINZ CENTER 2000:49).

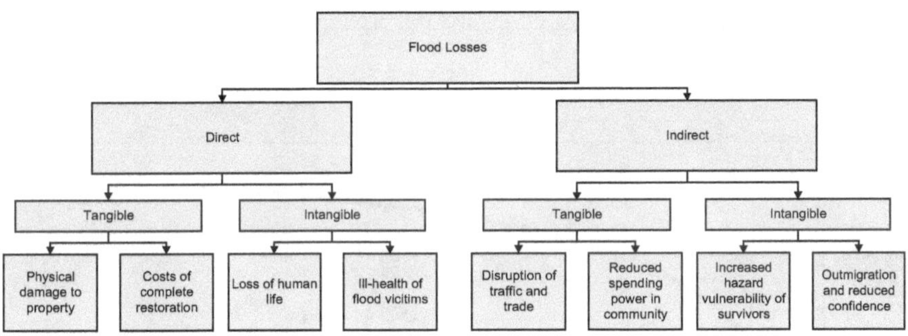

Figure 12: Categories of damage potential (Author's design, based on SMITH and WARD 1998:35)

Furthermore, the losses can be divided into tangible or intangible damage, whether they are measurable in monetary terms or not. Both subcategories can be divided into primary and secondary losses. Primary losses result solely from the event itself, while secondary losses follow the primary losses, for example expenses resulting from the reconstruction of primary losses (SMITH and WARD 1998).

3.2.1.1 Direct losses

Direct losses are caused by the direct physical impact on property. Two main approaches can be identified in estimating the damage caused by the natural hazard. The first method is based on the collection of information after the event using field interviews and questionnaires. The advantage of this method is the direct linkage to the actual event giving accurate data. However, the data collection method is error-prone due to the fact that it relies on information from the affected population. Immediately after the event, most people tend to overestimate the amount of damage to their private or business property, while the exact extent of longer-term structural damage may not be obvious at this time either. After a couple of months, on the other hand, the interviewees might be unable to recall the details of the actual damage. For further adaptation, this data is used to calculate the damage in relation to the severity of the natural hazard and the potential loss. Relying on the experience of previous events, it allows an easy transfer to other areas where natural hazards have not yet occurred. In addition to the interviews, investigations by civil engineers are conducted which classify the damage of buildings to derive average damage ratios in relation to the characteristics of buildings and hazards. This method is more precise, since it is highly standardised and the data quality is consistent.

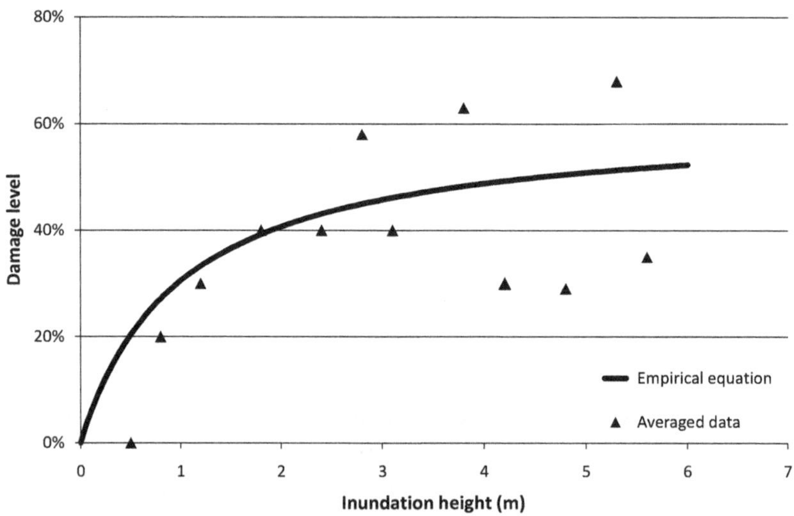

Figure 13: Average damage level as a function of inundation height (Author's design, based on RUANGRASSAMEE *et al. 2006:394)*

In the case of the 2004 tsunami, the location of the building in relation to the inundation depth can be identified as the major trigger for the damage caused (RUANGRASSAMEE et al. 2006:387). Based on the samples, the average damage level for the distance from the coastline and the inundation depth is calculated using regression techniques. By using statements of the population about their assets or interpolated secondary statistics, the possible amount of damage can be calculated and located in the inundation area. Another possibility to connect the depth-damage curves to monetary values is by using insurance claim data or secondary statistics about accumulation of assets in the area. In the next step, damage curves can be calculated that show the amount of damage in relation to the inundation height and the type of housing, allowing this knowledge to be adapted for other possible scenarios (cp. Figure 13).

This method can be easily adapted to different kinds of scenarios, but has a weakness in that it does not take into account mitigation measures applied by the population (SMITH and WARD 1998:39). The actual costs of the damage, i.e. the primary costs, might differ from the costs for the restoration after the event. These secondary costs tend to be higher due to the fact that after a natural hazard, a shortage of labour and materials often occurs (SMITH and WARD 1998:46). On the other hand, these repairs - and often also improvements - can increase the value of the houses in the long-run, especially in the case of singular events such as a tsunami (MONTZ 1992).

Another result of the direct impact of natural hazards is intangible losses that cannot be measured in monetary terms. Mortality, but also health problems resulting from physical or physiological suffering, could be a result of the impact of tsunamis. Findings from GUHA SAPIR et al. (2006) identified several groups displaying different mortality ratios among the population in Tamil Nadu in southern India (cp. Figure 14).

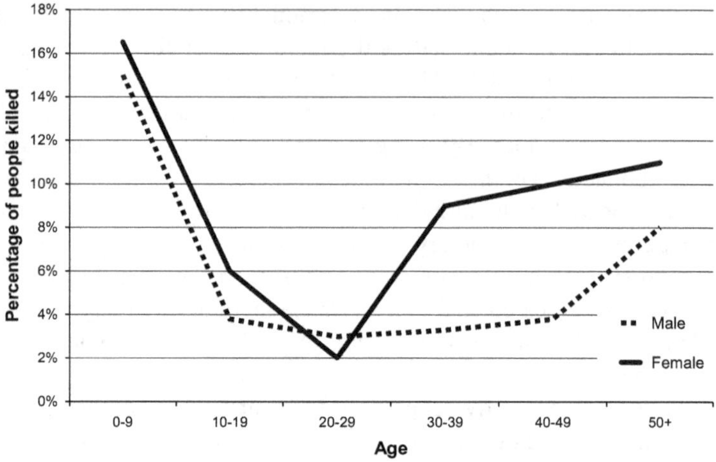

Figure 14: Proportion of deaths according to age of individuals (Author's design, based on GUHA SAPIR et al. 2006:31p)

Young children, the elderly and women between 15 and 50 in particular had the highest risk of death during the tsunami. People working in fisheries and their families were also highly at risk, since their workplace - and mostly also their homes - were located near the coastline (cp. Figure 15).

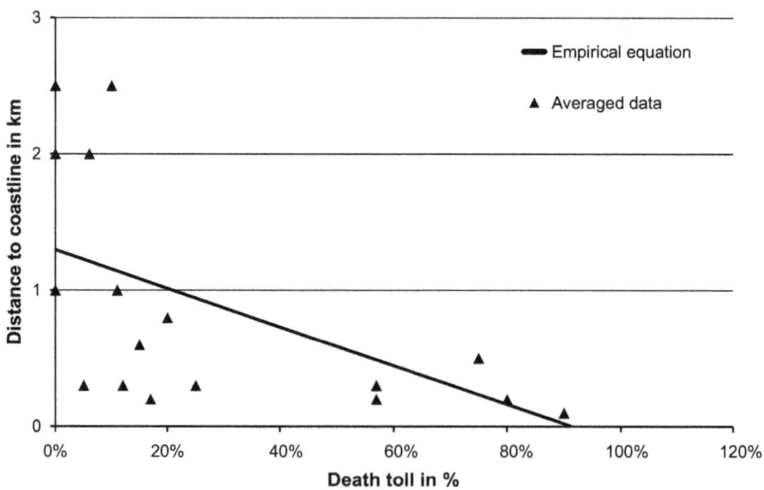

Figure 15: Proportion of deaths according to distance from coastline (Author's design, based on KATHIRESAN and RAJENDRAN 2005:604)

Injuries were mostly caused by debris floating in the water, but hygienic problems after the event also caused a secondary impact upon the population. After the event, many people suffered from psychological and mental problems caused by the emotional stress and the traumata.

3.2.1.2 Indirect losses

The impact of natural hazards is not only limited to the inundation area. As a consequence of the direct impact of the hazard occurrence, economic and social activities can be interrupted causing major economic losses. These indirect effects can affect households and companies not suffering directly from the advent of the natural hazard, but also those that have already been affected by the direct impact. These effects result from the relations within the economy. Interruptions of these relations due to damages and a decline or complete stop in production can cause severe losses for the suppliers (backward linked) or the sub purchasers and final consumers (forward linked) (cp. Figure 16).

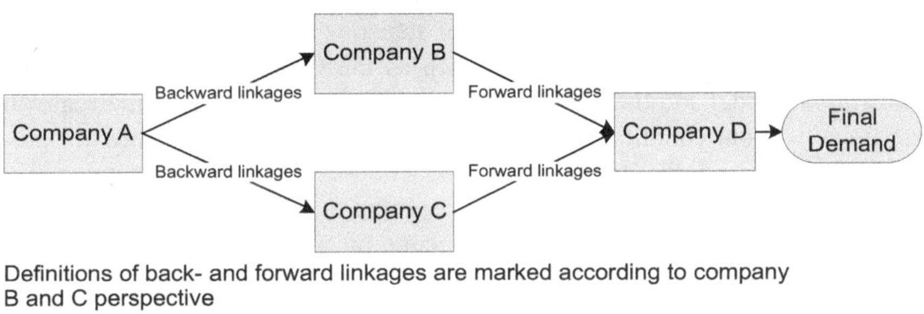

Figure 16: Economic interrelations causing indirect effects (Author's design, based on COCHRANE *2003:7)*

These effects can spread throughout the whole economy as a result of the multiplier effect influencing all parts of the economic system, i.e. the households, companies and the government in their demand or supply of goods, labour and money (WOHLTMANN 2007:87pp). These effects can be caused either by the interruption of traffic flows caused by the damage to infrastructure, the destruction of companies, the breakdown of final demand from the households or the shortage of labour after the event. According to VAN DER VEEN (2003:7), the magnitude of the indirect effects depends on the availability of alternative sources for supply and demand that can substitute the original ones, the duration of the disruption, which leaves the possibility of compensation for the losses by falling back on repositories, and the possibility for other, unaffected companies to extend their production. For the calculation of these indirect effects, the input-output tables of the System of National Accounts can be used. These tables provide information about the linkages of the business sectors and the multiplier effects deriving from these linkages (OKUYAMA 2003). The spatial effects on areas not directly affected are evaluated using a two-region input-output table covering the affected area on the one side and the rest of the country on the other side. By doing so, the indirect effects of a natural hazard are calculated for the years following the event, including the negative impacts resulting from damage and interruption of flows and the positive impacts of the reconstruction. This approach relies on the existence of a regionalised input-output table that fully specifies interregional trade flows according to region of origin and destination and to industry source and purchasing sector. Mostly, this data is not readily available at this detailed level of specification. Therefore, detailed and comprehensive surveys among the companies, households and governmental agencies are necessary in order to collect data about purchase and sales according to sector, both within and outside the target area. Several methods have been developed to by-pass this both time and finance-intensive survey-based regionalisation procedure by using secondary official statistics (SIM et al. 2007:2). ISARD (1998:87pp) describes several models, such as the Chenery-Moses model, the intra-national model and the gravity model, which aim to calculate interregional trade flows without conducting additional surveys. Published statistics are predominantly used to derive the regional production and demand capacities to regionalise the national input-output table, assuming that the national input-output coefficients can be used for the regional economy without taking into account the region's special production and demand structure. Other approaches use hybrid methods conducting surveys in the regions' leading industries and using non-survey methods for the other industries. One recent example of this hybrid approach is the MultiREG model developed for the nine federal states of Austria, including 32 sectors and four final demand categories. In a perennial research project, the national input-output table of Austria was combined with interviews of companies and wholesale trade companies throughout the country to derive the regionalised input-output tables. The model is supposed to simulate the regional and sectorial effects of investments in the federal states of Austria.

In Asia, only a few regional input-output tables have been compiled for Japan and China; in South-east Asia, only Vietnam possesses an interregional input-output table between Ho Chi Minh City and the rest of the country (TRINH et al. 2005:3). For Thailand, a bi-national input-output table has been compiled for the linkages between the province of Mukdahan in the north-eastern part of the Kingdom and the bordering province of Savannakhet in Laos (SIM et al. 2007). This table has been compiled on the basis of national input-output coefficients regionalised only with secondary statistics about the regional production and consumption. The authors assume that the technology of production in the province leads to the same coefficients as in the whole country, and that goods are 'imported' from other regions only when their demand cannot be satisfied in the study region. As a result of the inconsistent and differing quality of the secondary data, the revision of the regionalised input-output table needed 'intuitive judgement' for the adjustments (SIM et al. 2007:35).

Another method for the rapid assessment of the indirect effects of natural hazards is suggested in the guidelines of the Economic Commission for Latin America and the Caribbean (ECLAC). This method is especially aimed at regions with a 'prevailing insufficiency of reliable quantitative information' (ECLAC 2003:8). Because of this lack of secondary data, ECLAC suggests a more survey-based approach in combination with available secondary data in the affected areas. This survey should cover information about the direct damages of the natural hazards, the different types of input for the production and also the linkages to suppliers and customers in order to estimate the indirect effects on employees or other regions that might be affected by the decrease in supply (ECLAC 2003:206). If available, the input coefficients for the production process can be estimated by using secondary statistics from the government or other agencies, for example the input-output tables (ECLAC 2003:319). Based on the ratios and the direct damage, the spread of the indirect effects to other sectors or to the household's income can be estimated (ECLAC 2003:328).

3.2.2 Physical sensitivity according to the risk analysis

For the prediction of the damage to property in relation to the exposed assets, a location factor is necessary. This factor must explain the amount of damage which might occur in relation to the exposed elements. The characteristics of the wave on the coastline are influenced by a variety of indicators ranging from the geomorphologic to the biologic configuration of the coastline, as well as the exact location of the household or company. The wave energy is related to the duration and amount of water moved by the source of the tsunami, for example an earthquake or landslide, and the distance the water travels from the starting point of the tsunami. CHATENOUX and PEDUZZI (2007) used the indicator 'Distance from subduction fault line' for this purpose, which showed the highest beta value in their study predicting the run-up height of the tsunami waves in several countries bordering the Indian Ocean. Due to the fact that in the TRAIT Project, all study areas are located within an area of around 100 km ranging from north to south, this indicator can be skipped. If the results of this study are to be transferred to a national or multinational level, the distance to the subduction fault line should be included to cover the differences in the amount of water moved and the velocity of the waves hitting the coastline. Impacts of tsunami waves generated near the coastline are usually greater than from those which have their origin further away from the coastline due to the loss of energy to friction or spreading (NOAA 2007).

On the micro-level, coastline parameters such as the change in water depth near the coast, the angle of the tsunami to the coastline or environmental components reducing the energy of the wave before hitting the coastline needed to be considered. As geomorphologic indicators, CHATENOUX and PEDUZZI (2007) tested the influence of several indicators on the run-up height of tsunamis. The indicator 'coastline angle with the tsunami wave' indicates whether the coastline is exposed directly to the wave or the wave is running parallel to the coastline. The 'coastal elevation' was measured at a distance of 5 and 30 meters to estimate the slope of the coastline. Measuring the depth of the near-shore bathymetry, the 'length and slope of the proximal slope' were calculated and included in the model. A steep slope generally blocks the energy from the waves, while a flatter slope builds up higher waves. This is represented in the multiple-regression model by a negative sign before the coefficient of the 'average depth at 10 km'. The same influence is generated by the length of the slope showing a positive coefficient sign, indicating that a longer slope generates higher run-up tsunamis than a steep one. In a second step, environmental features were included in the study. The study encompassed coral reefs, sea grass, mangroves and different types of land cover. The results show that the distance behind sea grass beds was shorter than in other areas, while the presence of coral reefs shows an opposite effect (cp. Table 2).

Table 2: Parameter combinations to predict Tsunami run-up heights (Author's own design, based on CHATENOUX and PEDUZZI 2007:296)

Variables	B	p-level
Distance from subduction fault line	-0.83	0.00
Average depth at 10 km	-0.31	0.01
Length of proximal slope	0.64	0.00
Percentage of sea grass beds	-0.13	0.00
Percentage of coral	0.16	0.00
Intercept	8.70	0.00
$R^2 = 0.655$ N = 56		

The four study sites chosen in Thailand are only partially located behind sea grass beds or mangrove areas, so these indicators have not been included in the study (cp. Map 1).

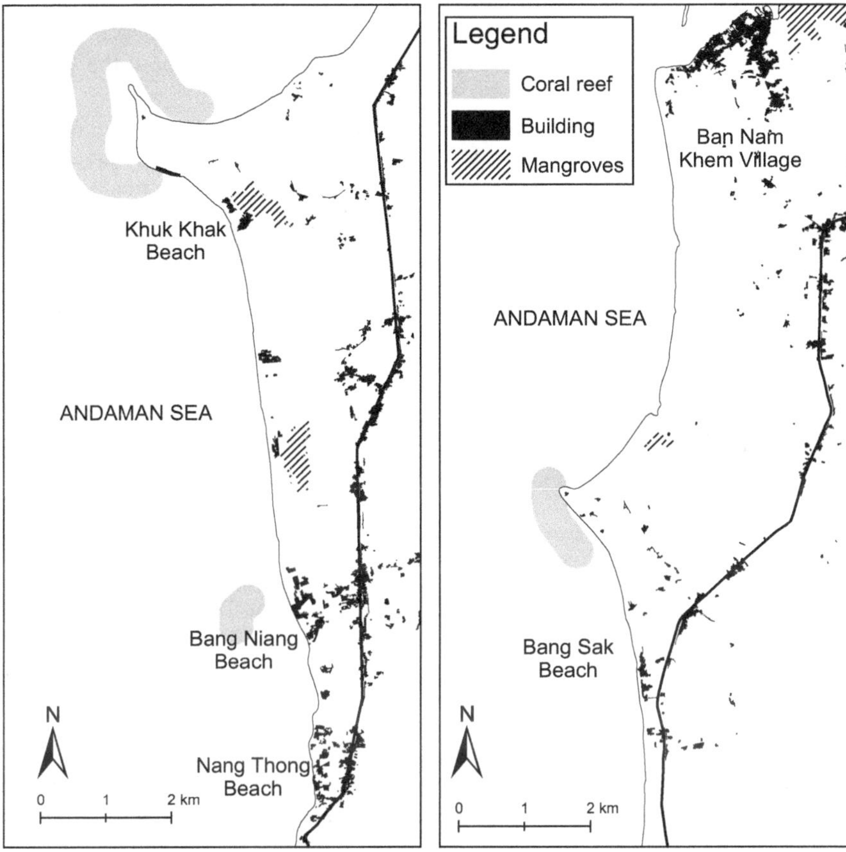

Map 1: Coral reefs and mangroves in study area in Phang-Nga (Author's own design)

For this study, the bathymetry data provided by the United States National Oceanic and Atmospheric Administration (NOAA) was used. The ETOPO1 dataset is a 1 arc-minute global relief model of the Earth`s surface, integrating land topography and ocean bathymetry (AMANTE and EAKINS 2009). ETOPO1 Data was used to calculate the length of the proximal slope and the depth at 10 km from the coastline. Information about the location of households and companies results from the Digital Terrain Model (DTM) providing information about the height above sea level with a resolution of 15 cm. This DTM was derived from pictures taken with a high resolution stereo camera developed by the German Aerospace Centre (DLR). This stereoscopic camera contains nine image sensors taking pictures of the earth's surface from five different angles.

This stereoscopic disparity enables the derivation of information about the elevation of each point of the surface and also offers true-colour ortho-photos (DLR 2004). From this dataset, information has been derived about the location of each household and company, including the height above sea level and the distance from the coastline. VERMAAT and THAMPANYA (2006:2) found a significant negative correlation between the distance of households from the coastline and the loss of property in fishing villages in Tamil Nadu, India. The indicator 'distance from coastline' accounted for an R^2-value of 0.59 and showed a significant impact. Their study used a non-linear regression model, but also tested linear regression, which showed no significant difference. Another indicator affecting the impact of the tsunami is the 'Wave Heights' of the 2004 tsunami. These ranged from six metres near Thai Muang to 12 metres in Khao Lak (IOUALALEN et al. 2007:4).

Table 3: Parameters of physical sensitivity for study communities (Own calculations, based on TRAIT Household Survey; AMANTE and EAKINS 2009; TRAIT MFC Data)

	Average depth at 10 km in meter	**Max. wave height in meter**	**Distance to coastline in meter**	**Height above sea level in meter**
Ban Nam Khem	-17.4	11.0	698.0	4.6
Thai Muang	-34.3	6.0	1,873.1	9.3
Khao Lak	-32.8	12.0	1,633.5	12.9
Patong Beach	-32.6	8.5	926.8	10.9

For the prediction of the actual loss of the households and companies caused by the tsunami in 2004 these indicators will be tested for their influence on the outcome of the tsunami in the study areas for households and companies.

3.2.3 Economic sensitivity deriving from regional structure

Household and company vulnerability is strongly related to the industry structure that dominates the area as well as to the individual's source of income or the business sector in which the company is involved. Indirect effects in particular lead to a slower recovery of households and companies from the direct impact of an external shock (CUTTER et al. 2003:247; MILETI 1999). In the relevant literature, several indicators are discussed describing the vulnerability of the economic structure in a region. Vulnerability analysis on the household level often integrates the share of income from the primary sector. It is assumed that business sectors depending on the extraction of resources are especially prone to the impact of natural hazards based on their location (HEINZ CENTER 2000:120pp; CUTTER et al. 2003:253). In addition to this, these activities are often carried out by self-employed individuals who lack capital for the substitution of assets and have no access to foreign investments (EAKIN 2005). Indicators determining the vulnerability of the primary sector can also be connected to other business sectors. FLAX et al. (2002:167) refer to the necessity of analysing the special economic situation in each region for its major economic sectors and also its regional distribution. This includes the identification of the businesses that are most important to the community in terms of employment, tax revenue or supply, i.e. companies that might be affected by the loss of supply from business sectors located in risky areas. In a second step, businesses which are located in risky areas must be identified. Several authors refer to the positive influence of the inflow of external resources, whether as an inflow of external spending power or foreign direct investments (NAUDÉ et al. 2007; LIOU and DING 2004). It is argued that these foreign direct investments allow a faster recovery, since the headquarters are less likely to be affected in the advent of a natural hazard. In the case of dependency on the inflow of external spending power, positive results can be found for export-oriented businesses (ATKINS et al. 2000), while tourism-oriented businesses tend to be more vulnerable (CALGARO and LLOYD 2008).

Covariant natural hazards such as the tsunami affect a wide area of coastal land. It is not only households and companies which suffer from the consequences of this impact, but also public and private infrastructure located in this area. Besides the economic vulnerability, it is commonly acknowledged that the dimension of physical vulnerability also affects people's livelihoods (MECHLER 2004:19). Physical vulnerability refers to the susceptibility to damage of engineering structures, such as roads, lifelines (electricity, telephone or sewage systems), or facilities, such as schools. The interruption of this infrastructure affects the livelihoods of households located in the area in several ways. Besides causing a handicap in everyday life, the economic situation might be affected by hindrances to people arriving at their workplaces, which might lead to a reduction in income. It can also affect households depending on remittances from migrated household members by interrupting the banking system (DESHINGKAR and AHEEYAR 2006). This can also happen to households not directly and physically affected by the natural hazard.

3.2.4 Elements of household vulnerability

Resilience capabilities of households can be divided into individual and institutional possibilities and characteristics on which the household can rely in the advent of disaster (BILLING and MADENGRUBER 2006:405). Individual strategies refer to strategies resulting from the household's individual or internal capacities and potentials for dealing with the consequences of natural hazards. Institutional resilience capacities are provided by external institutions such as the government or NGOs, or other agencies which provide financial or non-financial support.

One of the key elements in household vulnerability is the access to human capital as mentioned in the sustainable livelihood framework (WHITE et al. 2004). Human capital encompasses skills and knowledge on which the household can rely in the case of the occurrence of an external shock. SCHULTZ (1975) argues that educated persons can adapt more easily and flexibly than other individuals to a change in the economic circumstances because they use assets more efficiently and exploit new income possibilities. As a consequence, these households should be able to recover more quickly and adapt to the new situation. Furthermore, better educated households are more likely to understand early warning information and to access recovery help (HEINZ CENTER 2000). The stability of employment of households is in relation to their education level. More highly educated employees tend to be more valuable for their companies and are therefore less likely to lose their jobs in the case of an external shock. As a consequence, the influence of a labour contract on the likelihood of reduced income in the aftermath of the natural hazard should be evaluated in addition to the education level. The regional element of real income mentioned by SCHÄTZL (2003) can be related to the household's income, which allows the household to absorb losses more quickly and to extend its resilience capabilities in the aftermath of the disaster (CUTTER et al. 2003; BLAIKIE et al. 2007; ZOU and WEI 2010).

The size of the household might also affect its vulnerability. For families with a large number of children or elderly people depending on limited financial resources, the vulnerability could be greater. On the other hand, having a larger number of income sources can decrease the household's vulnerability due to income diversification and a higher income level (CUTTER et al. 2003; BLAIKIE ET AL. 2007).

Several studies have shown that households which can rely on more income sources have a higher capability to cope with external shocks (DEJANVRY et al. 1991; KINSEY 1998). There are varying degrees of income diversification that have to be evaluated differently. While diversification of income by engaging in part-time jobs or unskilled labour helps to increase the household's income, it is not suitable for extending the coping capacity (BARRETT et al. 2001:324). Therefore, using only the number of income sources of a household is not a suitable indicator for measuring income diversification in relation to an extended coping capacity. In this context, income diversification has to be measured by including the number of income sources and their relative importance for the total household income. To measure the degree of diversification for the household income, the Shannon equitability index is a suitable tool, including the number of income sources as well as their share of the total income (SCHWARZE 2004:28p). It is calculated by multiplying the share of each income i by its logarithm and summing up all income sources S of each household. In this study, business sectors have been used to distinguish between the income sources of the households:

$$H_{income} = -\sum_{i=1}^{S} [(incshare_i) * \ln(incshare_i)]$$

The measure H_{income} is calculated for every household and increases with greater diversity. To normalise this measure on a scale of 0 to 100, the Shannon equitability index E is calculated using the formula:

$$E = \left[\frac{H_{income}}{-\sum_{i=1}^{S}\left[\frac{1}{S} * \ln(\frac{1}{S})\right]}\right] * 100$$

The denominator is the theoretical maximum possible Shannon index if the household were to use all available income sources. Therefore, the Shannon equitability index ranges from zero to 100 and gives the percentage share of the household's income diversification in relation to the maximum possible diversity of income. According to the statistical standards used by the National Statistical Office of Thailand, 16 business sectors were identified and a further 12 other income sources, including transfer payments from relatives or governmental agencies. A special role is played by remittances sent to the household as a form of in-kind payment (PELLING and UITTO 2001:56). CLARKE and WALLSTEN (2003) studied the effect of remittances in the aftermath of a hurricane in Jamaica, finding that the amount of remittances increased in relation to the damage the household suffered. They also found that the remittances act as a form of unofficial insurance, even if they could not cover the complete damage to the household's property.

Despite the fact that diversification is highly acknowledged as one of the key elements of risk management at the household level, SIEGEL and ALWANG (1999:23) refer to the possible outcomes of a diversification strategy. They argue that possible outcomes of this strategy might include increased return and lower instability, but depending on the particular household circumstances, all other combinations, including decreased return and higher instability, are possible. Households with farm activities might diversify their plantings in several ways, but for households with limited land, a likely outcome of this strategy is reduced instability but less return, since the market prices for small suppliers tend to be lower than for farmers offering a larger number of crops (ALWANG et al. 1996). In terms of off-farm employment, households also follow the strategy of income diversification. This often includes the involvement in small businesses or seasonal mitigation arrangements. REARDON et al. (1988) documented these strategies particularly for poorer households. SIEGEL and ALWANG (1999) argue that these strategies can be effective in lowering risk, but are mostly followed by lower returns and increased costs. As a consequence, they might be suitable as second best strategies due to low efficiency compared with external and formal insurance arrangements.

The affected households' and companies' own financial resources play an important role in the recovery process. If these resources are available, they allow an immediate reaction to the physical damage, and therefore a faster completion of the whole process, whereas households depending on external sources need to wait until donations are available or risk management processes are initiated. Households' or companies' own resources may result not only from savings or loans in the formal banking system, but also from insurance claims (ALDERMAN and PAXSON 1992). These formal arrangements require functioning market institutions encompassing the banking and insurance system or other private formal organisations such as NGOs, which cover this field with arrangements such as microfinance or microcredit systems. Informal arrangements are often used if there are no functioning market institutions or if the access to these institutions is not available, but a mix of formal and informal coping strategies can also be found (HEITZMANN et al. 2002). Informal arrangements for risk coping include financial transfers beyond formal arrangements, such as borrowing from friends or family. Intra-community transfers also play an important role if not all households in one area are affected equally (HOLZMANN and JøRGENSEN 2001). These arrangements are open to all kinds of households, but might be limited in effectiveness and are more expensive in terms of direct and opportunity costs (MORDUCH 1999).

In cases where internal resilience capabilities are not sufficiently available, affected households are not able to compensate for the external shock or need to rely on external capabilities. For instance, these can be provided in terms of financial support or by donations from governmental or non-governmental agencies. BENSON, TWIGG and MYERS (2001) evaluated the role of NGOs in the risk reduction process. They distinguished between post-disaster actions, undertaken to minimise the impact of a disaster that had just occurred, and coping activities and long-term hazard risk reduction measures that help to increase the preparedness by reducing the impact or strengthening the internal resilience capabilities. The study identified major groups of long-term risk reduction efforts such as disaster preparedness training for local disaster management agencies, renaturation of environmentally degraded areas, establishing microfinance systems, and, to a minor degree, undertaking research and advocacy initiatives. As an immediate reaction to disasters in the short term, the study identified fields such as the provision of materials or resources to assist the recovery process, the provision of food and first aid equipment, or the construction of houses. These tasks are not only undertaken by NGOs, but also by governmental organisations involved in the risk management process.

The framework for providing external support to the affected population can be taken from the sustainable livelihood framework. This argues that a livelihood comprises the capabilities, assets (materials and social resources) and activities that are required for the means of living. A livelihood is only sustainable if it can cope with and recover from external shocks (SCOONES 1998:14). Therefore, it is crucial to provide emergency relief during the first weeks and months following the disaster, including the support of survival equipment and tools. When the humanitarian needs are covered, reconstruction and development aid, supporting services and incentives should be provided to enable the population to establish their livelihoods in a sustainable way. RÉGNIER et al. (2008) identified a clear distinction within the recovery process in India and Indonesia in the aftermath of the tsunami of 2004. The first approach is to initiate a large-scale reconstruction of infrastructure and housing. Besides the reconstruction of the facilities, its aim is to stimulate multiplier and spillover effects both up and down the value chain to restart the local economy. The second approach concentrates more on the microeconomic sector by stimulating entrepreneurial activities in various sectors to encourage the local economic activities in this way. Reconstruction support differs in terms of local participation, time schedule and financial investments from external agencies. A showcase for the different approaches used in the recovery process is the reconstruction of houses. DUYNE

BARENSTEIN (2006:2) divided the approaches pursued by the NGOs according to the participation of the local population. On the one hand there is the owner-driven approach, where the external source only delivers building material and knowledge support for the inhabitants to restore their houses themselves. This approach is suitable if there is enough labour force available and houses are constructed relatively easily. On the other hand, there is the contractor-driven village reconstruction, where the new houses are constructed in a different place to reduce costs and to decrease the exposure to natural hazards. Here, no participation of inhabitants is included and the houses are prepared completely by an external contractor. The study showed that people living in owner-driven-reconstructed houses are more satisfied with the process than people living in houses built by contractors that are not familiar with the local situation. Economic aspects of the different approaches used by the NGOs have not been evaluated in this study. Comparing the approaches to the differentiation made by RÉGNIER et al. (2008) clearly shows that only approaches integrating the local economy are suitable to allow a fast recovery of the whole economic system in the affected region.

Table 4 provides a summary of the previously discussed elements of household vulnerability deriving from the three different perspectives of vulnerability based on previous results from empirical studies on households in emerging countries such as Thailand. These elements were sorted according to the previous definition of exposure, which encompasses a household's economic assets (whether physically available or cash flows), its sensitivity defined by characteristics leading to a perturbation of the exposed elements, and its resilience perspective, which consists of elements allowing a household to cope with this impact and to return to a stable and acceptable situation.

Table 4: Potential indicators for household vulnerability (Author's own design)

Perspective of vulnerability	Indicator	Vulnerability is expected to be:
Exposure	Household assets	--
	Income level	--
Sensitivity	Location of household i.e. closer to sea, lower above sea level	Increased
	Devastation of regional infrastructure	Increased
	Higher devastation of infrastructure	Increased
	Employment in primary sector	Increased
	Employment in business sector dominating the region	Increased
Resilience	Income level	Lowered
	Education	Lowered
	Size of household	Undefined outcome
	Possibility of risk transfer (insurance, ext. recovery capabilities)	Lowered
	Secure employment situation	Lowered
	Income diversification	Lowered
	External support	Lowered
	Possibility of usage of own, family or friends financial resources	Lowered
	Access to banking system	Lowered

3.2.5 Elements of business vulnerability

Despite the fact that the amount of research on the vulnerability of businesses to environmental disasters is small compared to the huge number of studies on household vulnerability, it is assumed that business vulnerability also depends on the resource level, owner characteristics and other business criteria (WEBB et al. 2002; ZHANG et al. 2009). Therefore, this approach can also be traced back in its roots to the entitlement and sustainable livelihood theories discussed above (cp. Chapter 2).

On the side of the companies, the exposed assets are determined by the company's book values of fixed assets. These can be separated into four different categories, namely building and construction, machinery, vehicles, and other equipment such as office appliances (NSO 2007b:57). Cash and securities must also be counted among the exposed assets. These are often less vulnerable to natural hazards, since they are processed electronically and are not affected by the physical impact of a natural hazard. In addition, a disaster can cut the linkages to suppliers if they are affected by the hazard and force the company to adjust to material shortage, even if it is not affected by the hazard itself. This 'domino' effect on production chains can cause an economic multiplier effect (NRC 1999:37pp). As well as the loss of suppliers, businesses can also be affected indirectly by losing customers because of population casualties, dislocation or the absence of customer groups. Businesses serving a market that is limited to the disaster area face this problem more than others (NRC 1999:37pp; WEBB et al. 2002; cp. Figure 17).

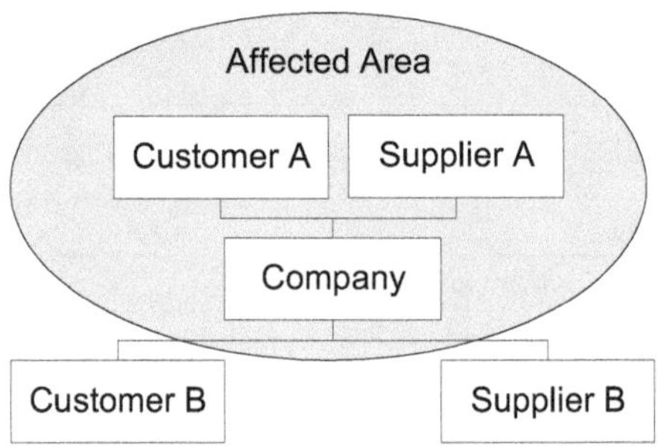

Figure 17: Company relations in the disaster area (Provided by author)

This element of sensitivity particularly affects small retail companies who face the loss of nearly all their sales caused by short-term customer loss. Businesses serving supra-regional or international markets experience less impact on their sales (KROLL et al. 1991). It is not only direct customer relations which are prone to interruptions: inter-business connections might also be affected by natural hazards, but under less vulnerable conditions, since these connections often tend not to be focused on a small region. Following this line of argumentation, it can be seen that a mono-structured sales and supplier dependency leads to a higher vulnerability level in the case of the loss of a supplier or customer group (ATKINS et al. 2000; ZOU and WEI 2010). Therefore, the actual number and share of the customer groups is compared to the maximum possible number of customer groups, including people from the area, tourists and/or companies. In the same way, the supply structure of the companies is evaluated and integrated into the statistical model. Supra-regional or international market-serving companies, however, are not exempt from indirect effects either, since interferences of business operations can result from the interruption of vital production input or the interruption of transport to customers. Besides the interruption of supply goods, constraints can also result from employee casualties, such as deaths, injuries or illness, the temporary dislocation of employees, or the preoccupation of employees with the restoration of their livelihoods. Furthermore, employees not directly affected by the hazard can be kept from their workplace by the interruption of traffic infrastructure. It is particularly businesses who cannot replace their employees easily, since they depend on a highly skilled or trained workforce, which face the results of this effect. In addition to the breakdown of customers or suppliers, indirect effects can be caused by the interruptions of infrastructure, such as water, electricity, transport and telecommunications, which might force businesses to shut down after the disaster and prevent employees from accessing their workplace (WEBB et al. 2002; TIERNIEY and NIGG 1995). TIERNEY (1997) found that as a consequence of the Northridge earthquake, the loss of lifelines was one of the main reasons for business closure. In their study on the recovery process of businesses in southern California, TIERNEY and NIGG (1995) found that the time of closure after the hazard had a significant influence on the probability of the business being able to recover after the event. This results from a reorientation of the former customers in order to satisfy their demand, which might be covered by other companies in the meantime (ZHANG et al. 2009). After the earthquake in Kobe in 1995, the port of the city suffered from a significant reduction in cargo handling, even though the physical damage had already been repaired. One reason for this was a reorientation of logistics companies towards other ports in the region (CHANG 2000).

Sensitivity towards indirect effects is not only determined by regional factors such as infrastructure or the degree to which employees suffer from the impact, but also by company-internal factors such as the business sector the company is involved in. After an external shock, the demand structure usually shifts away from consumer goods towards the effort to restore the damaged buildings. Therefore, relevant industries such as construction, building materials and home furnishings can experience a short-term growth in sales (NRC 1999:45pp). If the demand caused by the damage is satisfied, an abrupt drop of demand in this sector in the long-run is often experienced (ZHANG et al. 2007:47).

In the case of natural hazards, economic sectors that rely on the use of natural resources are especially exposed to the impact, i.e. the agricultural and fishing sector. Besides the physical damage to business property, these sectors are also prone to changes in their farming areas or fishing grounds. The tourism sector also relies on the natural potential of a region, with 'consumer confidence, reputation and positive imagery' being an even more crucial element (GURTNER 2007:218). As a consequence, tourists simply change their holiday destination if there is a disaster in the destination area. This depends mainly on their perception of the area, for example whether they feel unsafe or associate a negative image with their planned destination (SONMEZ et al. 1999; SUH 2004). Tourism is widely recognised as a helpful starting point for regional development and the reduction of poverty (GOODWIN 2007; WTO 2005). In the early stages of this development, tourism often has a special importance for the region, since it is a singular income source and people's livelihoods strongly depend on this one source alone (KNOX and MARSTON 2008:381pp;). This dependency also spreads to back-linked business sectors such as tourism-related retail or tour companies. Therefore, business sectors depending on tourism are especially vulnerable (CHAIPAN and CHAIPAN 2006). In the long run, the negative image transported in the media coverage of the disaster can lead to a vicious circle by reducing the consumer demand and interest in the region. This can lead to airline carriers withdrawing from the affected location and inhibiting the rapid recovery of the affected sector (GURTNER 2007:228). To capture the effects deriving from the special situation in a specific business sector, WEBB et al. (2002) include dummy variables for three business sectors, namely wholesale/retail, manufacturing and service. In the final OLS regression model, only wholesale/retail showed a significant influence in the explanation of the recovery time in one of the two areas surveyed. On the company side, it must be assessed as to which business sectors contribute to the regional economic situation in terms of supply and attracting income flows into the region.

The elements of companies' internal characteristics determining their coping capacity result from the entitlement theory. Previous research on the survival of companies indicates a variety of firm-level factors leading to a lower sensitivity to external shocks and a higher resilience capacity. It is commonly accepted that the age of the business plays an important role in its ability to cope with external shocks. The direction of the influence therefore seems to be a controversial issue. The majority of authors refer to a higher probability of survival for older companies (YASUDA 2005). In their study on the recovery of companies after natural hazards, WEBB et al. (2002) found that younger organisations have a lower probability of failing when they have to face an external shock or the consequences of a natural hazard. BARRON et al. (1994:414) refer to this phenomenon as 'liability of obsolescence', meaning that older companies have a higher risk of failure in the case of external shocks due to organisational inertia and a lower capability to adapt to new situations. Another criterion influencing the sensitivity and resilience of companies is the size of the firm. In their study on business recovery after the Northridge earthquake, DAHLHAMER and TIERNEY (1998) found that larger firms are significantly more likely to recover within one and a half years than smaller companies. This may result in a causal relation that larger firms usually have more financial backup than smaller firms and are often part of a chain, national or even multinational firm. Also, their activities are more spread out on a national or even international market than those of small companies. Micro-enterprises with less than four employees or informal business activities tend to have a particularly high vulnerability.

In the research on household vulnerability, it is widely accepted that external help provided by governmental and non-governmental agencies leads to a faster recovery after an external shock (BLAIKIE et al. 2007). This might also be the case for companies affected by natural hazards. Therefore, it is crucial to test the effect of the amount of aid received by the company and the number of sources that provide external help. External help might be supporting the rebuilding and restoring process in the businesses themselves, and stimulating supply and demand within the region. Besides this, national and international economic trends in the specific business sector or its linked sectors are also likely to restart the economy within the affected area and allow a faster recovery process (DAHLHAMMER 1998).

Table 5: Potential indicators for company vulnerability (Author's design)

Perspective of vulnerability	Indicator	Vulnerability is expected to be:
Exposure	Fixed assets	--
	Sales volume	--
Sensitivity	Critical business sector i.e. tourism or agriculture	Increased
	Focus of sales on tsunami-affected region	Increased
	Focus of sales on tourists	Increased
	Higher devastation of infrastructure	Increased
	Location of company i.e. closer to sea, lower above sea level	Increased
Resilience	Supply diversification	Lowered
	Reconstruction-related products	Lowered
	Sales diversification	Lowered
	Level of sales	Lowered
	Education level of employees	Lowered
	External support	Lowered
	Multi-unit business	Lowered
	Days of closure after hazard	Increased
	Business age	Undefined outcome
	Positive economic trend in business sector	Lowered
	Number of employees	Undefined outcome
	Access to banking system	Lowered

3.3 Determining the level of vulnerability

Because it is commonly agreed that vulnerability relates to the consequences of a perturbation on the affected system, it is necessary to take a closer look at the process following the impact of a stressor (ZSCHAU and KÜPPERS 2003:330). Therefore, TWIGG and others argue that vulnerability is often counter-posed with capacity to anticipate, cope with, resist and recover from hazard impacts (TWIGG 2004).

CARTER et al. (2004) differentiate between three stages in the economic anatomy of an environmental shock, which they define according to the time elapsed and the asset stock of the affected system (cp. Figure 18). The first stage and the starting point of the outcome of a vulnerable status is the shock itself, destroying physical assets and causing a reduced income or sales volume as a consequence of indirect effects. After the shock itself comes the coping period, in which the affected household or company starts to deal with the consequences of the impact. To do so, it resorts to its specific coping capacities. CARTER et al. (2004) refer to the examples of borrowing financial resources from formal or informal sources, increasing the working time to generate additional income, cutting back on consumption, or selling remaining assets. This is followed by the recovery period in which the assets are restored and the household returns to a stable situation. The time needed to return to this stable status depends on the household's sensitivity and its resilience capabilities. The boundaries between the different phases are blurred, but the final point of the recovery period can be precisely captured using statements from the household or company about whether they feel recovered.

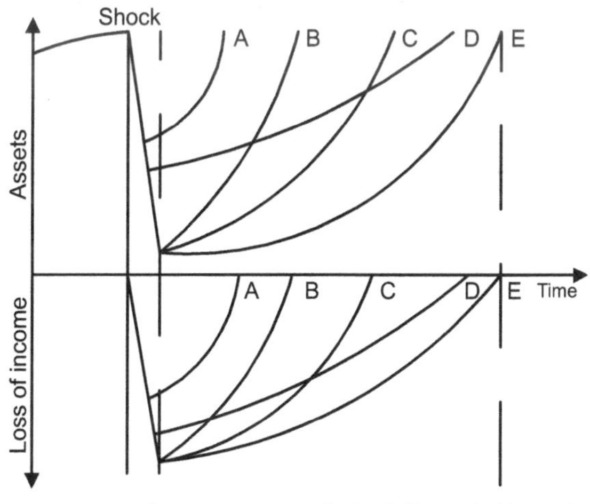

Figure 18: Economic anatomy of an environmental shock (Provided by author)

Therefore, it is obvious that the duration of the recovery processes depends on the amount of physical damage, the indirect effects and the coping and/or resilience capacity of the affected household or company. This was also recognised by WISNER (2002), who pointed out, for example, that difficulties arising during the recovery process can be understood as an aspect or an outcome of vulnerability. Based on this idea, BIRKMANN et al. (2006) developed a 'recovery index' for tsunami-affected households in the Sri Lankan cities of Galle and Batticaloa. This index includes the damage incurred by the household and its disposable income that can be used to finance the restoration. In a survey, the damage to the household was classified and the estimated reconstruction costs were calculated. These costs were compared to the available income of the household. From this data, it was calculated how long the household needs to finance the restoration process. The advantage of this method is that it provides a measurement tool for estimating the potential recovery time directly after the impact in relation to the damage and the income achieved by the household. Another application would be any ex-ante evaluation of the vulnerability status. On the other hand, other socio-economic factors that might contribute to a faster recovery, and therefore reduce the effects of the hazard, are not considered. This applies to all factors that cannot be linked to the income situation, such as support from external sources, i.e. from agencies or social networks. (BIRKMANN and FERNANDO 2008:101).

In their study on the vulnerability of community businesses to environmental disasters, ZHANG et al. (2009) used the time companies needed to reach the pre-disaster sales level to identify different vulnerability groups of companies. The authors identified four different groups of companies according to their recovery after the advent of a natural hazard (cp. Figure 19).

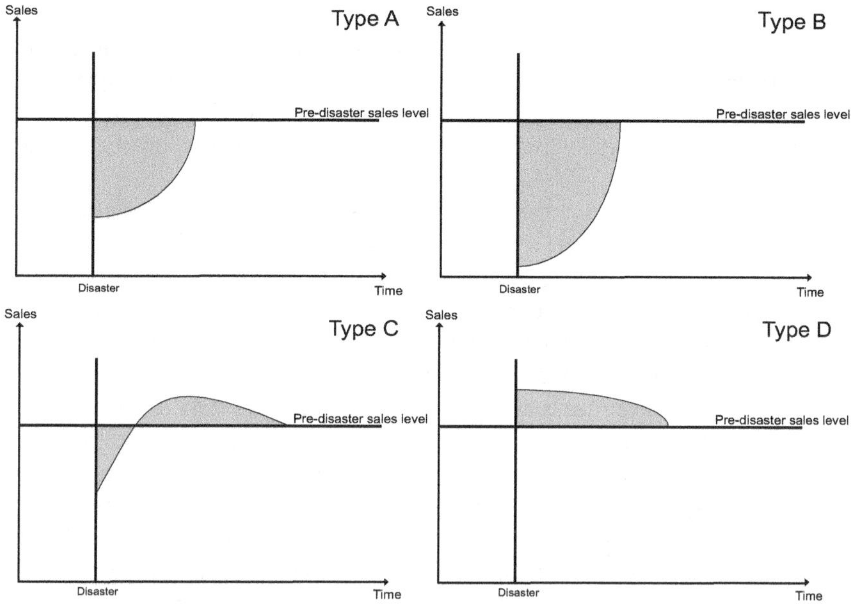

Figure 19: Idealised recovery process of company sales after a disaster (Author's design, based on Zhang et. al. 2009:47)

Their classification is mainly bound to the business sector in which the company is involved. They distinguished between two major groups of effects following the occurrence of natural hazards. The first group is formed by businesses inside the damaged area which only experience limited disaster impact and have high coping capacities, such as professional business services (Type A). These businesses tend to recover quickly and therefore only show a limited vulnerability level. On the other hand, there are businesses such as large manufacturers which experience larger initial decreases, leading to a longer recovery period (Type B). These might be caused by higher damages and a reduced coping capacity.

Besides negatively impacted companies, groups of businesses can also be identified which benefit from the impact of the environmental disaster through an increase in their sales volume. Businesses located inside the affected areas, but which are only slightly affected in terms of physical damage, might be hindered in their business activities due to breakdowns of infrastructure for a shorter time. These companies experience a short reduction in sales, but can benefit after overcoming this problem from the fact that their competitors might not be able to satisfy the demands of their customers (Type C). Those who particularly benefit from the impact include non-affected businesses near or within the affected areas which offer recovery-related products for construction, distribution or production of construction material, and the hospitality sector providing hotel rooms or restaurant services for construction workers, volunteers or other officials. These businesses benefit from the special shift in demand and the breakdown of other companies, which causes a shortage in supply on the market (Type D).

Another approach used in the vulnerability analysis of businesses after Hurricane Andrew and the Loma Prieta earthquake in 1989 near San Francisco analyses the recovery in a more comprehensive indicator, taking into account the number of employees, the number of customers served, business profits and the owners' impressions about the general business situation, bringing them together in a combined index (WEBB et al. 2002). This index was derived from data measuring the recovery process eight years after the Loma Prieta earthquake and six years after Hurricane Andrew. Therefore, it is only evaluated as to whether the business had recovered at this specific time or not. Information about the point of time at which the business finished its recovery process is not included.

From the ex-post perspective of vulnerability assessment, it is possible to investigate the complete process of recovery with the re-accumulation of assets that were destroyed by the natural disaster and the recovery of the income or sales over the time period. The actual recovery time as a measurement index for vulnerability integrates all sensitivity and coping capacities on which the household or company could rely during the handling period after the event. This makes it possible to derive groups of households or companies with specific characteristics and similar recovery periods for future applications. The point of being recovered is determined by the perception of the affected household or company. DACY et al. (1969:71) argued as early as the 1960s that hazard-affected residents desire to be at least in the same economic position following the disaster as they were before. This can imply that the community reaches the same economic level as before, but can also grow further than the pre-disaster economic level. The discussion about the recovery process was initiated by CLINTON in his proposition about ‘Building back better’, demanding that good recovery must leave communities safer by reducing their risk and by increasing their resilience capabilities (CLINTON 2006). This follows the older definition mentioned above and can be translated into economic terms in that the economic situation in the affected area should reach a new equilibrium similar to the one before or exceeding it (ROSE and GUHA 2004). Therefore, the opinion and perceptions of the residents can reflect when the recovery process is completed. FOLKE (2006) refers to this approach, which understands resilience as a counterpart of exposure and sensitivity as the engineering resilience which focuses on the idea that a system aiming to recover a status of a stable equilibrium, and therefore vulnerability, can be measured in return time and the efficiency with which this process is organised. But surely vulnerability and hazard impacts also include opportunities to develop new risk management strategies by recombining structures, processes and characteristics of a system. According to FOLKE, this research strand is still in an explorative phase, and therefore not measurable in terms of modelling. It will be accessed in this study using qualitative research methods (FOLKE 2006:263).

4 Data base of the study

For the assessment of vulnerability in the affected region, quantitative and qualitative data-were combined. This is due to the fact that this study pursues two major aims: on the one hand, the assessment and quantification of the vulnerability situation triggered by the tsunami of 2004, and on the other hand, that of the recovery period which followed. For the assessment of vulnerability and recovery after the tsunami of 2004, a quantitative, standardised questionnaire was used. The majority of the questions had been designed as closed questions, with open questions rarely being used for gathering statements concerning the recovery process. This design was chosen in accordance with the purpose of the study, i.e. producing a quantitative and generic model based on a system of hypotheses derived from several studies on vulnerability theory (cp. Chapter 3).

Previous research and a consistent theoretical framework are necessary as a basis for applying a quantitative method including closed questions (DIEKMANN 2010:531). The main reason for choosing a qualitative and open approach to assessing the mitigation strategies of households and companies in the aftermath of the tsunami of 2004 was the specific characteristics of quantitative research methods. Three specific characteristics in particular play important roles: the focus on the research subject, i.e. the special perspective of the interviewee, the openness, i.e. open questions without giving predetermined answers, and the analysis in the familiar surroundings of the interviewee (DIEKMANN 2010; MAYRING 2002). This triangulation should support the verification of the quantitative generic model (BLOOR 2001:11).

4.1 Secondary data

Secondary data were used in this study especially to investigate the vulnerability situation of the affected regions on higher levels than the household or company level, and for the selection of sample sites. These data were provided by official governmental institutions, namely the National Statistical Office with its regional offices in the Phang-Nga and Phuket Provinces, the National Economic and Social Development Board, and the Bank of Thailand. Databases provided by these agencies are mainly available in printed form or, in limited numbers, via online access. In general, the statistical system of Thailand is well developed and broadly follows international standards according to the classification systems used in the publications. Metadata describing the sampling procedure, definitions used and aggregation procedures are usually included. Deficiencies can be found in the data collection, where the sample sizes are often constrained by budget limitations. Therefore, the statistics tend to be representative only on the level of provinces, and rarely for lower regional units such as Amphoes (administrative districts). Also, the base year for constant prices in the national accounts is 1988, which is hardly comparable with international statistics (IMF 2006:11).

- *Statistical Yearbook (NSO):* These publications are distributed annually by the regional offices of the NSO for each of the 75 provinces and for the whole kingdom. Compiled from data from various agencies, they contain information regarding population, economic activities and regionalised national account data, i.e. the gross provincial product.
- *Household Socio-Economic Survey (NSO):* This survey is published annually by the NSO, presenting information for the whole kingdom and for the four statistical regions. It includes information on economics and social matters of the household, including income and expenditures, assets and liabilities, structure of the household, housing characteristics and transfer payments. In the whole country, 52,000 sample households were interviewed in stratified samples with a distinction between rural and urban areas.
- *Business Trade and Service Survey, Industrial Census and Agricultural Census (NSO):* The business trade and service survey covers businesses with at least one person engaged in the wholesale and retail trade, hotels and restaurants, real estate activities, recreational and other services. The survey is conducted every two years. It offers information on the characteristics of businesses involved in this sector, their employees, and their income and expenditures. The survey is representative on the level of the five statistical regions of Thailand. This census provides information on the basic characteristics of the companies and their employees, but no information on the economic situation. Finally, the agricultural census covers the last business sector. It was most recently conducted in 2003 for the fifth time, interviewing approximately 5.8 million agriculture holders (NSO 2003). The data is representative at the regional level and is integrated and annually updated in the provincial statistical yearbooks.
- *Gross Regional and Provincial Product (NESDB):* Tables about the gross provincial product (GPP) are published every year by the NESDB, which is in charge of providing national account data. The GPP data is available for the 16 sectors corresponding to the International Standard Industrial Classification (ISIC).
- *Input-Output Tables (NESDB):* I-O tables are published every five years for the whole country. These are available from 16 to 180 sectors using the Thai Standard Industrial Classification Scheme (TSIC 2001) which is derived

from the ISIC (UNSTATS 2002). Regionalised I-O tables are not available for Thailand. In this case, the national table was used to evaluate the linkages (ECLAC 2003:319). The raw tables provided by the NESDB were evaluated using the software IRIOS (Interregional Input Output Software) developed by the Department of Economics and Econometrics at the University of Groningen (OOSTERHAVEN and STELDER 2007:2). The analysis of an input-output table allows the identification of the major drivers of a national economic system. By analysing the forward and backward linkages for each industry, key buyers and sellers within the production chain can be identified that might cause major indirect effects on other industries if they are directly affected by a natural hazard. The input-output model used in this analysis is based on the standard demand-driven Leontief model, which can be noted as: $X=(I-A)^{-1}f$, where x is an n element industry output vector, f is an n element industry final demand vector, A is an n*n input coefficient matrix and I is an n dimensional identity matrix. By investigating the backward and forward linkages, it is possible to identify the integration of a business sector within the economic system. Forward and backward linkages are calculated by excluding one industry from the input-output table and measuring the impacts of this withdrawal on the remaining sectors. Here, the backward linkages of one industry measure the degree of dependency of other industrial sectors on the supply of inputs from this specific sector. On the other hand, the forward linkages indicate the importance of one industry in terms of the interdependency in other industries' inputs. (STELDER et al. 2000:15). Backward and forward linkages do not indicate the importance of one industry according to their share of value added to the GDP, but rather in terms of their importance within the economic circle. In the case of vulnerability and exposure analysis, this point is of great importance, since sectors with high forward and backward linkages could be the origin of the indirect impacts on a variety of other companies, whereas the economic impact of less linked industries does not spread out that far to other companies. Therefore, industries with a high backward linkage can be seen as 'key industries' which purchase a significant proportion of the outputs of other industries (ZHANG and FELMINGHAM 2002:10).

4.2 Household and Company Survey

Household and company surveys were conducted using a standardised questionnaire containing mainly closed questions. Interviews were conducted between January and March 2009 in the four communities of Ban Nam Khem, Khao Lak and Thai Muang in the Phang-Nga Province, and Patong Beach in the Phuket Province (cp. Map 2).

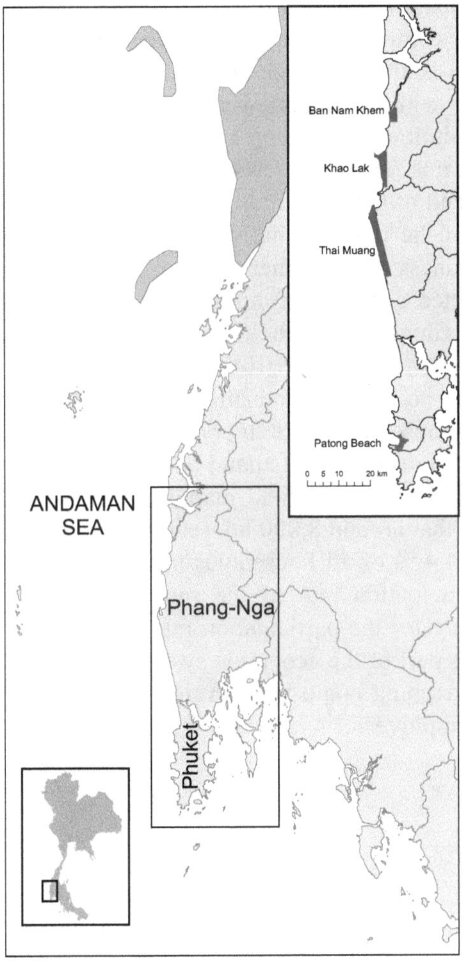

Map 2: Study regions in southern Thailand (Provided by author)

Face-to-face interviews were conducted by research assistants from the Social Science Institute at Chulalongkorn University. The interviewers were prepared by an introduction to the purposes of the study, the theoretical background, and the possible answers for each question along with its specific purpose within the framework of the questionnaire. Afterwards, the interviewers were trained in a role play situation among themselves to become more familiar with the questionnaire and the situation. Completed questionnaires were checked and irregularities discussed with the interviewer, and if necessary, the interviewee was contacted to complete the questionnaire. Data entry was carried out by the researcher for an additional check on data validity, and if necessary, unclear points were verified with the interviewer. After data entry, data was examined for errors and irregularities by running a descriptive analysis and checking scatter plots and histograms. The interviews were conducted from January to March 2009, which is the end of the tourism high season in this region. It was assumed that in this period of the year, the individuals and companies are less preoccupied with doing business than at other times of the year. Prior to the surveys, a pre-test was conducted covering 10 households and 12 companies, testing the questionnaire with around 5 % of the desired number of companies and households in the final survey (HÄDER 2010:387). After the pre-test, only minor adjustments were made to the questionnaires relating to the classification of answers. In total, 323 households were interviewed, covering 1,264 persons in the four affected regions in southern Thailand and 286 companies employing 2,215 persons (cp. Table 6). From the field mapping and remote sensing results, it could be concluded that around 8,000 households are located in the study areas, meaning that around 4 % of all households/inhabitants in the area could be interviewed. The participation rate of the sampled households reached nearly 75 %. A similar share for the participation rate could be achieved for the company sector. For this part of the economic system, nearly 15 % of all companies found in the field mapping could be interviewed, covering more than 10 % of the estimated total employees.

Table 6: Number of interviewed households and companies in the study areas (Author's design, based on TRAIT survey 2009)

Community	Number of interviewed households	Number of interviewed companies
Ban Nam Khem	104	39
Thai Muang	54	19
Khao Lak	66	64
Patong Beach	99	164
Total	323	286

The interview partner chosen was either the head of the household or another person who declared him/herself to be competent in answering questions about each of the household members and the tsunami impact on the household. In the case of company interviews, the owner or general manager was preferred as an interview partner. Because of the specific and detailed questions asked in the survey, three or more departments were often involved, and more than one person provided the necessary information about the business. Only in the case of rankings or questions about opinions was it ensured that the owner's or manager's statement was noted.

The two questionnaires were designed to cover all aspects of the economic situation in the surveyed region, both on the company side as well as on the household side, reflecting the hypotheses developed in Chapter 3.2. The household questionnaire comprised the following parts:

- **Section 1 - Localisation:** For the exact localisation of the household in the GIS system and the connections with the remote sensing picture, the coordinates were assessed, in addition to the address, using GPS devices. The exact localisation of the household is crucial for deriving information about the tsunami exposure of the household, namely its distance to the coastline, its height above sea level and other indicators.
- **Section 2 – Household members:** This section covered information about the occupation, income from wages and salaries, as well as other income sources and working hours of each household member. In addition, personal characteristics of each household member were noted, such as gender, age, marital status, education and absence from the household, for example migration. For the workplace of each household member, the working hours and location were noted in order to derive information about people's exposure for certain time slots, i.e. distribution of people during the night and day. In-

formation about the location of income source is also crucial for estimating possible indirect effects and mapping the interconnections within the economic system of the region.

- **Section 3 – Housing characteristics:** For the exact localisation of the interviewed household, information about the housing situation, such as type of house, building material, information about the roof and size of the house, was noted. These details help to locate the household within the GIS System, since GPS devices tend to have a margin of error of up to 30 metres, which can lead to incorrect allocations of households to buildings in remote sensing images, especially in densely populated areas such as Patong Beach. These indicators are also helpful for gathering basic information about the building stability and their resistance capacity in the case of a wave impact. Furthermore, building characteristics constitute helpful information for calculating the value of the house in combination with the rent for the house, or the calculated rent costs if the house and property are owned by a member of the household.
- **Section 4 – Exposed assets:** In addition to the property value, information about the possession of selected long-term consumable assets and vehicles was collected for the assessment of the exposed values of the household. Besides information about the price and the age of the item or vehicle, information about the location of the supplier was also noted. This information will be used to calculate possible positive indirect effects resulting from the recovery process in which damaged property needed to be replaced. Additionally, information about periodical expenses and the location where they are spent was gathered in order to assess the indirect effects in the case of loss of sales for the businesses connected to the household. Since not all long-term consumable assets existing in the household could be assessed because of time and budget limitations, only the information about selected items could been asked for. Therefore, items were chosen according to information from the socio-economic survey, determined by their rate of market penetration among the different expenditure groups of households. For example, air-conditioning units exist in the majority of high-income/high-expenditure households, but are almost non-existent in middle-expenditure households. At the other end of the range, mobile phones are available in nearly every household in southern Thailand (NSO 2007e:159). With this information, it is possible to estimate the total amount of assets for all groups of households.
- **Section 5 – Financial status:** The financial status of a household, especially its credits and savings, is a major influence for its coping capacity, allowing it to recover faster from damage to property and to absorb short-term income

volatilities. In addition to this, information about insurance was gathered, which allows the household to transfer its risks to an external institution.

- **Section 6 – Tsunami 2004:** This section assessed the outcome of the tsunami of 2004 for the household in terms of direct effects, i.e. damage to property, and indirect effects, i.e. loss of income, unemployment or personal casualties for each household member, and effects of damage to infrastructure on the household's livelihood. Besides the effects, coping mechanisms were also investigated separately for each shock, namely for the duration of the recovery period and the source of finance for repairing the direct damages, or occupational strategies that were undertaken to minimise the results of the indirect effects on the working situation.
- **Section 7 – Adaptation strategies**: To estimate the lessons learned from the tsunami for each household, the level of threat from tsunamis felt by the household was estimated.
- **Section 8 - Socio-cultural adaptation:** To develop a basis for the qualitative focus group discussions, information about the long-term impact of the tsunami on the community was gathered, and a short evaluation of the resilience strategies used in the aftermath of the tsunami by institutions such as the NGOs and the government was requested. This information was used as a starting point for the focus group discussion, as it painted an initial picture of the mitigation and adaptation strategies implemented by the households.

Parallel to the interviews of affected households, a second group of research assistants from Chulalongkorn University interviewed companies affected by the tsunami using a questionnaire comprising similar topics to the one developed for households. As much as possible, the structure of the household questionnaire was retained. In the following presentation of the questionnaire, only differences to the household questionnaire will be discussed:

- **Section 1 – Localisation:** Coordinates and address of the business were noted in order to localise the company in the GIS for further derivation of location indicators from remote sensing data.
- **Section 2 – Characterisation of business:** General characteristics of the business include information about the foundation year, the structure of the company, i.e. single unit, branch of domestic or international company, external investments and the structure of the sales portfolio. This follows the hypothesis that older companies are more likely to survive external shocks, since they have more experience, stronger connections to their customers and often greater financial resources. Branches of companies or companies with external investors have the advantage that their headquarters are located outside the inundation area and financial resources can be received. Finally, in-

formation about the product portfolio reveals information about the companies' dependency on special markets and their diversification, which can increase the resilience in the case of the collapse of single markets. Information about the patterns in the market in which the business operates was also examined, assuming that, for example, businesses operating in a market with seasonal or cyclical changes are dependent on savings from periods with high sales, and which might not be sufficient to bridge longer times with a decline of sales.

- **Section 3 – Business property characteristics:** Parallel to the household questionnaire, information about the business property was assessed in order to increase the accuracy of the localisation of the company in the Geographical Information System and to derive the property value.
- **Section 4 – Employees:** The number of employees and the amount of wages paid to the workforce allow a conclusion about the indirect effects sent out from the interviewed company in the case of a business interruption towards the households. It can also be tested as to whether the size of companies leads to a higher or lower resilience potential. Here, the education in particular might have a strong influence on the resilience capacity, since it is argued that companies which offer highly specialised and knowledge-intensive products are less likely to lose their customer base in the case of a temporal business interruption. Additionally, the typical working hours of the company were noted in order to derive information about the number of people at risk at different times of the day.
- **Section 5 – Assets:** For the estimation of the damage potential in the case of the direct impact of natural hazards, the amount of investment assets in the company was noted. According to the official statistics of the NSO, the interviewee was asked to estimate the value and the age for the groups of equipment/tools, vehicles, machinery/office equipment, and other inventory. For the estimation of interregional linkages and indirect effects of the tsunami, the location of the supplier was also assessed.
- **Section 6 – Financial situation:** Corresponding to the information requested in the household questionnaire, questions were asked in this section about savings, credits and liabilities which can be used in the case of the direct or indirect impacts of external shocks in order to cope with the effects. Additionally, questions were asked about the availability of insurance which might cover the effects of natural hazards.
- **Section 7 – Linkages to other companies:** This section contained information about the interrelations of the interviewed company with their suppliers and customers. In order to draw a picture of the economic connections within the affected region and to other parts of the country, the location, the

periods of time and the average amount of purchases or sales was noted. This allows an estimation of the indirect effects on other companies linked to the affected companies in the inundation area.

- **Section 8 – Effects of the tsunami:** Direct and indirect effects of the tsunami of 2004 for the company were assessed. For the direct damage to company property, the same categories were used as in section five, asking for the amount of damage, the amount of money invested in repairs and the length of time needed. Indirect effects on the company were assessed by asking the interviewee to estimate the percentage of business activity after the tsunami and the duration of time needed to reach the previous level of business activity. Additionally, the causal factor for the indirect effect from the point of view of the interviewee was asked about, i.e. supply, sales, employees, or the infrastructure.
- **Section 9 – Adaptation:** The final part of the questionnaire was designed to provide a starting point for the focus group discussion on adaptation strategies developed by the companies or on other levels relevant to the company. Therefore, the interviewee was asked to rank the threat of a tsunami for its business in the future and to give a short statement about strategies developed by his/her company.

The study communities are located in the parts of southern Thailand which are economically very attractive due to various job opportunities in the tourism sector and a significantly higher income level than other regions, especially compared to the north-eastern region (cp. Chapter 6.1). The official records on the number and address of the population are not reliable, since the law on residential notification requirement is not handled strictly in the Kingdom of Thailand (KRAAS 1996). Therefore, the official records could not be used for determining the basic population of the study area. Correspondingly, the socio-economic household survey by the National Statistical Office of Thailand does not rely on official population registers, but uses a sampling procedure based on random choice within selected blocks in cities or within small villages (NSO 2007c). In order to assess the size of the basic population in the study, a field mapping was conducted, recording information about building characteristics and the usage of the buildings. The field mapping was carried out by fifteen students from the Institute of Economic and Cultural Geography from the University of Hanover on a field trip to the region in September 2008. Based on this information, a stratified sampling plan was developed covering two percent of the total population of households and companies. These strata are subsets that share common characteristics (GROVES 2009:113). This plan is oriented towards the business sector of the company on the one hand, and on the other hand towards households based on the type of building in which the household lives, which was the main factor influencing the damage the household suffered during the tsunami of 2004. Additionally, this approach allows further investigations using remote sensing and transferring the interviewed household sample to the basic population of households. These can be identified using an automatic building classification. A representative sample for business sectors and housing condition in the region could thus be achieved (cp. Table 24).

4.3 Focus group discussion

In addition to the quantitative approach using a standardised questionnaire to assess the vulnerability and the recovery process following the tsunami of 2004, qualitative studies were used to verify the elements of the recovery process integrated into the statistical modelling and to identify the adaptation strategies that were developed later. In focus groups, a specific research topic is discussed in a small group of six to ten people under the supervision of a trained moderator. The discussion takes around 60 to 90 minutes (KEPPER 2008:162pp). The moderator initiates the discussion, integrates participants and focuses the discussion on the original discussion topic. The discussion is taped for further analysis afterwards and minutes are also taken. The interaction within the group is the most important characteristic of this empirical research method, since people respond to each other's contributions (CAMERON 2005:157). This interaction within the groups leads to a process in which the participants explore different ideas and opinions and, as a result, reconsider their own ideas and understandings. Therefore, the discussion process in a group can initiate an uninhibited environment that can stimulate the participants to express their opinions more freely. As a consequence, statements by the group participants can emerge in a focus group that would not be expressed in personal interviews, where no social interaction takes place. On the other hand, the interaction within the group involves the problem that the participants orientate their expression according to the group bias or individual opinion leaders, with the result that not all statements are expressed. Therefore, it is crucial that the moderator encourages all persons to participate in the discussion and to express their opinions, even if they are opposed to the group norms (KAMENZ 2001:112; BEREKOVEN et al. 2009:91). The task of the moderator is it to draw out the range of views on the research topic, to encourage disagreement within the group and to create a consolidated opinion formation (MYERS 1998; BLOOR 2001).

Focus groups have been used in geography and social sciences for a variety of purposes. As a singular method of collecting information, it is suitable for providing insights into people's opinions that might not be revealed by other qualitative methods. ZEIGLER et al. (1996) used focus group discussions to assess people's perception of the risk management during and after Hurricane Andrew. Their findings showed that the technique was suitable for gathering an insight into the people's beliefs that were not accessible using face-to-face interview techniques. At the end of the research, they were able to provide recommendations for the improvement of risk management based on these findings. Besides the use as a single data gathering tool, a supplementary use in combination with other methods is also possible. Integrated within a multi-method mix, outcomes of focus group discussions can be used to supplement quantitative research. For example, as a first entry in a research topic generating theories and research questions, focus groups can serve according to their nature as a qualitative research method, providing explorative knowledge. Focus group results can also be helpful following quantitative studies. Providing deeper insights into people's beliefs, the results can supplement the findings of standardised surveys and help to interpret the statistical results contesting or qualifying them in relation to the commonly agreed statements of the focus group participants (BLOOR 2001). Furthermore, focus groups can serve as a platform for presenting initial research results from previous surveys as introduction statements initiating the discussion. This introduction statement by the moderator allows the researcher to receive early feedback on his initial statistical analysis and gives the participants chance to comment on the findings. Positive feedback can also be seen as a further validation of the outcomes of the research (DOUGLAS 1976:131). Despite this positive outcome, WYNNE (1996) refers to the problem of 'reflective capacity' among the participants of focus group discussions who might know how the data is used and try to influence the outcome of the research. For example, in discussions related to previous risk management, participants can attempt to devalue the outcomes of the recovery process in the hope of gaining personal advantage, such as the extension of financial support.

For the success of focus group methods, the group compilation is a crucial point. It should be considered that all participants of the focus group are from a 'natural group' that relate them to the topic under discussion. Depending on the research topic, the group can be compiled by choosing participants with homogeneous or heterogeneous characteristics (CAMERON 2005:161). Composing homogeneous groups means that the characteristics of the participants within the group are consistent, for example in terms of age, economic status or social status. To achieve results for a variety of groups, multiple focus groups can be arranged covering different perspectives and participant groups. Homogeneous groups have the advantage of allowing easier discussions, since all participants have a similar background. More controversial discussions are encouraged by participants with different backgrounds. This can promote the discussion within the group and might lead to more open statements. The discussion is initiated by a short introduction to the topic by the moderator. This might be an introductory controversial statement, a film or the presentation of previous research results. The purpose of this introduction is to provide an overview of the topic and to initiate the discussion. After that, the discussion is supported by the moderator with questions or controversial statements. In the tradition of qualitative research, the session is based on guiding topics which help to structure the discussion but do not have to be mentioned in a specific order (CAMERON 2005:164).

The major aim of the application of the focus group was to identify the economic recovery and the mitigation strategies derived from the experience with the disaster. Therefore, the investigations concentrated on the major business sectors in the study regions in the Phang-Nga and Phuket Provinces, i.e. the agriculture and the tourism sector (cp. Chapter 6.1). In order to compose homogeneous groups, but also to cover a wide spectrum within the business sectors mentioned, three or four focus groups per major business sector were held. People involved in these businesses were invited to the focus groups in order to discuss this topic from the perspective of their experience in business as well as their personal livelihoods. The participants were chosen on the basis of the previous household and company survey. In total, 14 focus group discussions were held, involving 55 participants (cp. Table 7). Each of the participants was interviewed before the beginning of the group discussion to gather individual information about his or her occupational background, experience during the tsunami of 2004 and personal recovery process. The discussions were held in September 2009, which is the rainy season in Thailand and therefore a more quiet time for the agricultural and tourism sectors. Despite this, some focus groups could not take place due to short notice commitments of the participants. In general, focus groups should have at least four participants to achieve sufficient discussions (CAMERON 2005:162). In some cases, therefore, in-depth interviews were conducted as a substitute for the focus groups using the guiding questions developed for the focus groups.

Table 7: Participants of focus group discussions (Author's own design)

Agriculture		**Tourism**	
Background of participants	**Study area**	**Background of participants**	**Study area**
Plantations	Ban Nam Khem & Khao Lak (in-depth interview)	Hotels	Khao Lak & Patong
Fishing	Ban Nam Khem & Khao Lak	Restaurants	Khao Lak (in-depth interview) & Patong
Incidental Farmer	Ban Nam Khem & Khao Lak	Tour agencies and operators	Khao Lak & Patong
		Tourism-related retail	Khao Lak & Patong (in-depth interview)

As a starting question, the participants discussed the impact of the tsunami on their private and occupational livelihoods. The major topic of the discussion groups was the disaster preparedness and risk management strategies for the participants concerning their occupational and private life derived from their previous experience. According to the different players involved in risk management (cp. Chapter 2.4), the questions covered the household and company level, the community, the government and the non-governmental organisations. At the end, after discussing the different strategies, the participants were asked to identify constraints and effectiveness of the risk management approaches.

The discussion was moderated by a researcher from Chulalongkorn University, Bangkok. The moderator and her team are highly experienced in conducting focus groups. For the venue of the focus groups, community centres were chosen to create an informal setting which could easily be accessed by all the participants. In order to create a more relaxed surrounding, snacks and soft drinks were provided. As an entry for the discussion, a short presentation was given regarding the general effects of the tsunami of 2004 on the region and preliminary findings about the recovery process based on the household and company survey. This opened the discussion and provided some input for the discussion process. Since the majority of the focus group participants had also been interviewed in the household or company survey, which was conducted one year before, positive feedback was expressed regarding the commitment of the project in returning to the area. This was stated by several participants and possibly helped in obtaining more reliable answers, since trust between the research group and the local population had already been established. The discussion process took place in the Thai language. For further evaluations, the discussion groups were taped and parallel notes were taken containing statements and observations from the moderation team. Based on the notes and the tapes, reports were filed detailing the outcome of the discussions and the personal backgrounds of the participants. These reports were analysed using a coding scheme developed from the guiding questions and the experience and results gathered in the structural equation model, with the focus on the lessons learned from the tsunami in order to develop adaptation strategies concerning future natural hazards (BERTRAND et al. 1992). In keeping with the three perspectives of vulnerability, the coding scheme investigates the elements exposure, sensitivity and resilience mentioned by the participants:

Economic effects of the tsunami 2004
- Short-term effects of the 2004 tsunami impact
- Long-term effects of the 2004 tsunami impact

Sensitivity
- Conditions causing physical sensitivity
- Elements of the regional economic system influencing individual sensitivity
- Individual characteristics influencing sensitivity of company or household

Resilience
- Internal coping capacity on the household or company level
- Coping capacity resulting from external measures
- Adaptation strategies developed after the event

4.4 Critical examination of the survey methodology

The Boxing Day tsunami of 2004 attracted global media interest due to the large number of personal casualties and the extensive damage in coastal areas in South-east and South Asia as well as in parts of Africa. A major part of the devastated area is known globally as a tourism area, and in addition to the personal casualties among the local population, a large number of tourists from Europe, Australia and the Asia-pacific region were also affected. Media coverage led to enormous donations for national and international relief organisations and also to several research projects in the region. This strong focus led to several problems conducting empirical research in the region, even five years after the event. The households and companies interviewed for the TRAIT survey had often already participated in other surveys, and therefore sometimes refused the interview or expressed their mistrust in research because of previous experiences with researchers gathering their data without giving feedback to the community. Because the TRAIT research project involved intensive cooperation with Chulalongkorn University, which has a very good reputation in Thailand, with the Social Science Research Institute in particular having strong linkages to key persons and institutions in the region, a trusting interview atmosphere could be created. As a consequence, the participation rate in the household and company surveys was more than 75 %, which indicates the wide acceptance of the interviewers and the goals of the TRAIT project among the local population and businesses. The high degree of popularity and reputation of the researchers from Chulalongkorn University most probably also led to valid answers from the interviewees and prevented overestimation of damage or recovery times because it was made clear that the data would only be used for research purposes and not for political evaluation in order to increase the financial support for the region.

In their study on adaptive capacity in communities along the coast of the Andaman Sea, PATON et al. (2008) from the school of psychology refer to the special problem of conducting questionnaire-based survey studies in Thailand. Thai society is strongly characterised by its collectivist culture, which leads every individual to place a strong focus on the alignment with social norms and collective goals of the society, emphasising social relations and attempting to consider the beliefs of other people in planning their own actions (DIENER and SUH 2000:135). This leads to a significant problem in research using subjectively-oriented scales, for example the likert-scale, since people tend to adjust their own opinions according to the societal beliefs and norms, which might lead to lower scores. For example, people highly affected by the tsunami might reduce the description of their degree of suffering because of the fear of being called arrogant (BRISLIN 2000). This might lead to a lower variance in the data and therefore a less reliable analysis. In order to address this problem, this analysis of vulnerability and recovery relies on survey data assessed not as beliefs or opinions, but based on interval-scaled facts rather than underlying social beliefs or norms. Questions about opinions and beliefs were reduced to a minimum, and in the case of the section on mitigation strategies, used as a discussion base for the focus groups. In the focus groups, the statements and beliefs resulting from the survey on mitigation were discussed under the guidance of a moderator familiar with Thai society and western or individualist cultures, and therefore able to initiate a controversial discussion.

5 Methods of analysis

5.1 Structural equation modelling

As the primary analysis method, the structural equation modelling approach was chosen for this study, since it offers the opportunity to assess causal relations between theoretical constructs. The advantages and disadvantages of these approaches will be discussed in the following chapter. Basically, the method is a combination of different statistical methods which integrates path analysis and factor analysis into a single model. In contrast to exploratory factor analysis, factors and causal relations are specified by the researcher in advance, guided by theoretical assumptions and tested on a specific dataset (ARBUCKLE 2006:45pp). The model is designed by the user in the form of a scheme representing the hypothetical relations between latent variables and between the manifest variables and their latent variables (WOLD 1982).

In general, a structural equation model consists of an inner and an outer model (cp. Figure 20). The outer model specifies the relations between the constructs or latent variables and their explaining indicators. The inner model describes the relations between the latent variables. These can be divided into dependent latent variables respectively endogenous variables, and in independent latent variables respectively exogenous variables. In the following section, the latent variables will be named ξ (m=1… 1) and their related indicators $x_{(m)}$ j=1… j. Direction and power of the relations between the latent variables are represented by the path coefficients $\gamma_{(ml)}$. The connections between latent variables and their connected indicators are described as factor loading coefficients $\lambda_{(m)j}$ in formative models and as weight coefficients $\omega_{(m)j}$ in recursive models.

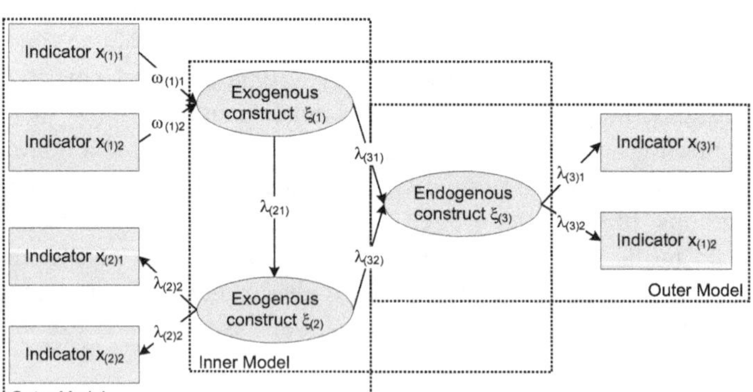

Figure 20: Exemplary path model (Provided by author)

Every relation in this SEM can be described by an equation. The interconnections between the latent variables in the path model can be described by a multiple regression model. In this model, the antecedent construct can be understood as the regressor and the following dependent construct as the sand: $\xi_m = \sum_{l \epsilon C^P_{(m)}} \xi_{(l)} * \gamma_{(ml)} + \mathrm{err}_{(m)}$. $C^P_{(m)}$ is the index of the antecedent constructs of the latent variables and $\mathrm{err}_{(m)}$ the error term. The outer model with the latent variables and their connected indicators is also explained by an equation. For reflective constructs, this is as follows: $x_{(m)j} = \xi_m * \lambda_{(m)j} + err_{(m)j}$. This equation includes the error term of the indicators. Formative models are represented by the equation resulting from the multiple regression models: $\xi_{(m)} = \sum_{j=1}^{J_m} x_{(m)j}\omega_{(m)j} + \mathrm{err}_{(m)}$ (BOLLEN 1989:16). The parameters of this equation will be estimated with either the partial least square approach or the maximum likelihood algorithm, which aim at reproducing the input data according to the specified model.

5.1.1 Methods of structural equation modelling

The model estimation procedures can be divided into two methods differentiated according to their use of covariance or variance as data input. Covariance-based approaches use the covariance matrix to determine the model estimators by trying to reproduce this matrix using the model specifications most accurate. This is done simultaneously for the whole model. By doing so, the approach aims for the reproduction of the relation between the indicators, but not as much for the reproduction of the case values. This shows that the covariance approach is best suited to the testing theories in their totality and for evaluating the interdependencies of indicators. By way of contrast to this, the highest aim of the variance-based approaches is the reproduction of the original raw data with the given constraints from the specified model (cp. Table 8). The covariance approach follows the idea of the factor analysis, which interprets the indicators as a linear combination of the factors. This is done by splitting the total variance of the factor in factor variance and error variance. Following this approach, the factors can be understood as the driving force behind the indicators explaining and causing their values. This helps to extract the part of the variance that is explained and caused by the factor. The principle component analysis used by the variance-based approaches attempts to represent as much variance as possible using linear transformation of the indicators on one principal component. This means that the variance of the indicators is not split up, but represented in total by the factor, and the loss of information is minimised As a result of this approach, the factors tend to be closer to the indicators and can be interpreted as more data-oriented, combining the indicators to one factor (WEIBER and MÜHL-

HAUS 2010:67). By eliminating the error variance, covariance-based approaches can be seen as more reliable for testing the relations in the structural equation model, while the variance-based approach attempts to reproduce the raw data given by the indicators and predict them more accurately. This shows that the covariance-based approach must be preferred if a study is aiming for model or theory testing, while a more data and prediction-driven study should choose the variance-based approach. Resulting from the step by step estimation used in the variance-based algorithm, the variance-based method tends to be more robust to small sample sizes. The holistic estimation of the covariance-based approach is more likely to produce misleading results showing high fit indices and/or path coefficients with a beta value higher than one - the so-called 'Heywood Cases' - if using small sample sizes (WEIBER and MÜHLHAUS 2010:159). Besides this, the variance-based approach does not rely on normally distributed raw data, while the maximum likelihood algorithm used in the covariance-based approach is based on the assumption of multivariate normal distribution.

Table 8: Characteristics of structural equation modelling approaches (Author's own design, based on WEIBER and MÜHLHAUS 2010:66)

Criteria	**Covariance-based**	**Variance-based**
Aim	Reproduction of covariance matrix	Prediction of data and variables
Theory relation	Theory-testing approach	Data and prediction-driven approach
Method	Factor analytic approach using a simultaneous estimation procedure of all parameter	Two-step regression-based approach for separate estimation of inner- and outer model
Aim of method	Minimisation of differences between empirical and model-based covariance	Minimisation of differences between estimated and observed data
Latent variable	Following the idea of factor analysis. Latent variable as a factor with isolated error term – latent variable as item-determining dimension	Dimension in the idea of principal component analysis – latent variable as item-combining dimension
Distribution of data	Multi-normal distribution	None
Quality criteria	Global and local goodness-of-fit criteria	Local goodness-of-fit criteria
Sample size	Larger samples	Smaller samples possible
Purpose	Confirmatory	Explorative
Software	LISREL, AMOS, EQS	PLSGraph, SmartPLS, LVPLS

Coming to a conclusion, the covariance-based approach tends to be more suitable for testing a complete system of theories and hypotheses. On the other hand, the variance-based approach is more qualified for predicting changes in dependent variables and for identifying the main determinants of a theoretical model, and is therefore used in this study.

5.1.2 Partial least square approach

5.1.2.1 Estimation procedure of the partial least square algorithm

The Partial Least Square (PLS) algorithm uses a variance-based approach. The aim is to minimise the variances in the error terms and the inner and outer models in order to reproduce the empirical data most accurately. Following this, LOHMÖLLER (1989:6) describes this approach as "... data-structure oriented ...". PLS uses a combination of various multiple regression analyses to estimate the parameters for the whole model in several steps. This algorithm was introduced by WOLD in 1957. The algorithm aims to minimise the error terms of all dependent variables by using various iterations. Dependent variables include the latent variables in formative constructs and all endogenous latent variables in the model. The indicators in reflective models can also be described as dependent variables.

The algorithm calculates an approximation of the parameters of the model in several iterations. In every iteration, weights and construct values are fitted to match the structure of the given path model, to minimise the residuals in the regression models and to fit the structure of the path model (HUBER et al. 2007:7). In three steps, the latent variables $\xi_{(m)}$ are estimated using the support of the environment variables $\xi^*_{(m)}$ and the weighting vectors $\omega_{(m)}$. These iterations are repeated until no more changes in latent variable values can be seen. When this test is positive, the inner model with its path coefficients is calculated and the outer weightings are estimated (cp. Figure 21).

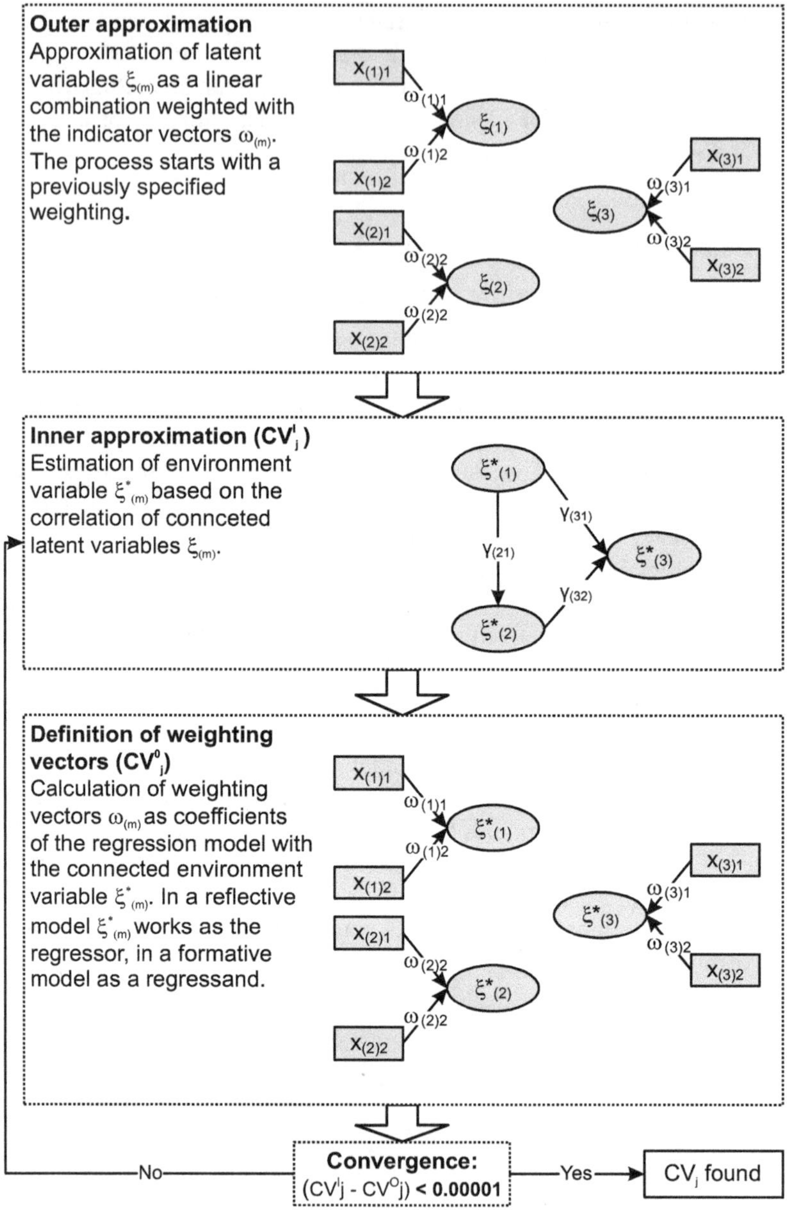

Figure 21: Iterative estimation of the partial least square algorithm (Provided by author, based on CONZE *2007:106)*

In the first step for estimating the latent variables $\xi_{(m)}$ they are estimated as a linear combination of their dedicated indicators. All indicators and variables are standardised by a mean of zero and a variance of one (CASSEL et al. 1999:438). Therefore, they are weighted with the vector $\omega_{(m)}$. In the first iteration, these vectors are set as a defined value and will be estimated in step three for further iterations. As an outcome of this step, a value can be estimated for each latent variable. In the first step, an outer estimation of the construct values is conducted.

In the second step, an environment variable $\xi^*_{(m)}$ is calculated for each of the latent variables $\xi_{(m)}$ using the weighted sum of the connected variables. In general, the algorithm uses the correlation coefficients of the antecedent latent variables (HERRMANN et al. 2006:38). The second step results in environment variables representing the relations between connected latent variables. In the third step, the outer weightings are estimated using the least square regression method and the resulting coefficients represent the weightings. In reflective constructs a simple multiple regression analysis is used with the indicators as independent and the factor as a dependent variable. In formative constructs, where the indicators are the dependent variables, a multiple regression with the estimated factor as a regressand is used. In the outcome this step improves the representation of the indicator values by the environment variable by minimising the residuals through the regression analysis. These three steps are continued as long as there are no more changes in the weights for the latent variables and convergence is reached (CHIN and NEWSTED 1999:316). Convergence is reached when the values of the latent variables estimated in the first step show only minor changes in comparison to the previous iteration. In this case, the final outer weights are estimated, as described in step three, and the path-coefficients among the latent variables are calculated using least-square regressions.

Since PLS is based on the principle of regression analysis, no assumptions of the distribution of the indicators is needed and a small sample size is also usable. As a guideline for sample size, CHIN (2009:311) suggests that the number of cases should be ten times the number of indicators of the latent variable with most indicators and the number of regression paths to the latent endogenous construct. Looking at the exemplary path model in Figure 20, a minimum number of twenty cases would be necessary.

5.1.2.2 Specification of the outer model

For the operationalisation of the measurement model, indicators are used which help to measure constructs that are not directly measurable. A distinction can be made between reflective and formative constructs. In a reflective construct model, the indicators are influenced by the factor. From this, it follows that the indicators result from the construct. In contrast to this, the indicators in a formative construct influence the construct. The formative construct can be understood as a linear combination of the indicators or as the dependent variable (WEIBER and MÜHLHAUS 2010:202). Following this idea, DIAMANTOPOULOS and WINKLHOFER (2001:271) understand the measurement concept of formative constructs as indices (cp. Figure 22).

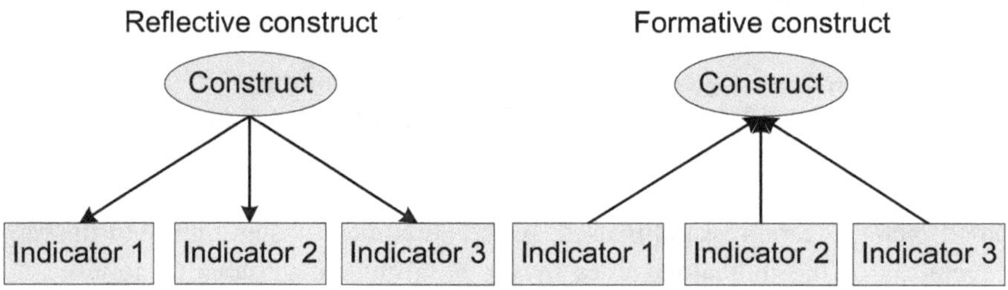

Figure 22: Reflective and formative constructs in the outer measurement model (Provided by author)

JARVIS et al. (2003) compiled a table of criteria to distinguish between formative and reflective constructs based on a meta-study of studies in the marketing field using structural equation modelling. Basically, three criteria were defined resulting from the causality of the relation between construct and indicators. These criteria need to be evaluated before specifying the model (cp. Table 9).

Table 9: Criteria for outer model specification (Author's own design, based on JARVIS et al. 2003:203)

Criteria	Reflective model	Formative model
Causal relation of indicator and construct	Indicators are manifestations of the construct	Indicators are defining characteristics of the construct
Mathematical relation of indicator and construct	Changes in the construct influence all related indicators	Changes in the construct result from changes in at least one of the indicators
Choice of indicators	Indicators should have similar content; dropping of indicators does not alter the domain of the construct. Indicators should correlate.	Indicators do not need to have the same content, dropping an indicator may change the domain of the construct. Indicators do not need to correlate.

For the construction of the factors, the approach of formative latent variables was used. The constructs in this model represent explanatory combinations of indicators determined by their combination of variables (FORNELL and BOOKSTEIN 1982). This means that the indicators define the factor, and a change in one of the indicators leads to a change of the factor. The indicators forming the factor do not have to be correlated, since the indicators stand for themselves and are not influenced by the factor. These factors are considered as index values representing their related indicators in the form of a linear combination (DIAMANTOPOULOS and WINKLHOFER 2001), whereas in reflective constructs, the indicators are understood as an expression of the factor itself (WEIBER and MÜHLHAUS 2010:36p). The idea of formative constructs, which can be understood parallel as index values, follows the idea of the theoretical framework of vulnerability with its different dimensions of economic systems (cp. Chapter 3). In addition to multi-indicator factors, some elements are only set up by one indicator, where no theoretically and statistically founded combinations can be formed. According to WANOUS and HUDY (2001), a combination of both factors in one model is acceptable and does not lead to limitations in the estimation process. Single-item constructs are especially useful for representing concrete constructs in contrast to the more abstract multi-item reflective constructs (NAGY 2010:85).

The goodness-of-fit criteria for formative constructs are defined by the way their parameters are estimated. Due to the application of the least-square regression method, a basic requirement is the absence of multi-collinearity, which is necessary to avoid linear interdependencies among the independent indicators. This correlation among the independent variables would lead to a biased estimation of the parameters (BROSIUS 2008:589). Therefore, indicators showing a high multi-collinearity need to be eliminated from the model. This can be tested using the variance inflation factor. For this test, one regression is performed for every indicator, using each indicator as an independent variable explained by the other indicators. The coefficient of determination of each regression analysis R^2 is used to calculate this VIF:

$$VIF_x = \frac{1}{1 - R_x^2}$$

This indicator represents the part of variance of indicator x which is explained by the other indicators of the construct. The VIF shows values starting from 1 and should not exceed 10 (BROSIUS 2008:569). Another constraint could arise from the occurrence of autocorrelation among residuals. As a consequence, significance tests return a biased and often suppressed result. Autocorrelation can be caused by the misspecification of the model due to the lack of relevant indicators. Additionally, autocorrelation can be caused if linear relations are assumed, but are in fact not present in the data (BROSIUS 2008:564). As a test for autocorrelation, the Durbin-Watson-Coefficient is widely accepted, which can range between zero and four. Values between one and three indicate the absence of autocorrelation (BROSIUS 2008:565).

Parallel to the regression analysis, PLS also provides t-values which can be used to check the significance of the independent variables. These are calculated using resampling procedures such as the bootstrapping or jackknifing algorithm, testing for the null hypothesis that the estimated path coefficients are not significantly different from zero. These generate subsamples from the total sample, which are used for the estimation process. By comparing the resulting parameter, the consistency of the estimation - and therefore the significance of the estimation - is calculated (BYRNE 2001:269). Relations based on a theoretically derived directional hypothesis can be observed using one-sided significance values based on the t-value (cp. Table 10).

Table 10: Goodness of fit for the outer model (Author's own design)

Criterion	Critical value
T-value	10 % > 1.28
	5 % > 1.65
	2.5 % > 1.96
	1 % > 2.33
Durbin-Watson-Coefficient	1 < DWC < 3
Variance inflation factor	VIF ≤ 10

DIMANTOPOULOS and WINKLHOFER (2001:271pp) suggest that the constructs should be specified according to theoretical assumptions underlining the content of the construct, followed by a statistical analysis, eliminating indicators as necessary, based on the statistical evidence provided by the variance inflation index or the significance tests.

5.1.2.3 Goodness of fit for the inner model

The PLS approach provides a variety of criteria for checking the goodness of fit of the inner model. In contrast to the covariance-based approaches, it does not provide global fit indices like the tests on parsimony or chi-square value indicating the closeness of fit between data and model (WEIBER and MÜHLHAUS 2010:172pp). Most of the criteria for the variance-based PLS approach result from the regression analysis and yield to judge the explanatory and prediction power of the model. As a standard, the algorithm presents the standardised path coefficients (Beta-coefficients) representing the effect of the independent on the dependent variable. LOHMÖLLER (1989:60) argues that a standardised regression coefficient higher than 0.1 can be seen as acceptable. Because the PLS approach does not rely on normally distributed raw data, parametrical significance tests cannot be applied, and the bootstrapping procedure needs to be used to judge the significance of the inner model parameter (cp. Chapter 5.1.2.1). The explanatory power of the exogenous latent variables on the endogenous latent variables is represented by the coefficient of determination (R^2) indicating the percentage of variance that is explained by the exogenous latent variables. Most of the software available does not provide the corrected R^2 by itself. This has to be calculated with the formula: $R_{corr}^2 = R^2 \frac{J(1-R^2)}{K-J-1}$ including the number of regressors (J), the sample size (K) and the degrees of freedom (K-J-1).

In addition to this, the prognostic effect of an exogenous latent variable on the endogenous latent variable can be identified by the effect size (f^2) showing whether there is a substantive impact. This f^2 value is calculated by the change in R^2 including the exogenous latent variable and by omitting it. CHIN (2009:317) suggests that f^2 values of 0.02, 0.15 and 0.35 can be identified as a small, medium or large effect (cp. Table 11). The effect size is calculated as:

$$f^2 = \frac{R^2_{included} - R^2_{excluded}}{1 - R^2_{included}}$$

Table 11: Goodness of fit for the inner model (Author's own design)

Criterion	Critical value
Standardised path-coefficient	≥ 0.1
f^2	$\geq 0.02 \geq 0.15 \geq 0.35$

5.1.3 Critical evaluation of structure equation modelling

Structural equation modelling is not widespread in vulnerability analysis, risk management or economic geography. Some authors use components of SEM in their studies. ATKINS et al. (2000), for example, used it in the first step factor analysis to estimate factor values which were used in a regression analysis in a second, separate step. As a dependent variable, income volatility of countries was used to create the commonwealth vulnerability index. An early application of an SEM model in the field of risk analysis was carried out by BURNS et al. (1993). Their study aimed to assess the public risk perception for natural hazards by modelling the relations between the physical outcomes of hazards, such as the amount of damaged property and the number of casualties, and the media's coverage of the hazard. Their model indicates an influence of both media coverage and physical outcome of hazards on the politics of improving risk management. This study can be counted as an early application of SEM in risk management, but it does not yet include elements of vulnerability analysis. Furthermore, not all current goodness-of-fit criteria, such as significance levels, were reported. The application of a full SEM model is new to vulnerability analysis, since the majority of studies use regression models or factor analysis, but do not combine these two methodologies and benefits from the possibilities of a full SEM. While the majority of applications are from the fields of business economics, sociology and psychology, few economic geographers or regional economists have used this approach in their studies. Examples from these scientific fields can be found in TAMÁSY (2005), who used causal analysis methods to identify determinants of entrepreneurial activities, EICKELPASCH et al. (2007), who transferred Porter's Cluster Theory to an SEM, and VAN OORT et al. (2009), who investigated the relations between urban knowledge economies and employment growth.

On the methodological side, one huge barrier for the application of covariance-based SEM is the demand for multivariate normally distributed data. Only if this specification is fulfilled can the maximum-likelihood estimator show consistent, unbiased and efficient estimates (PANTEN and BOßOW-THIES 2007:317). This criterion of multivariate normal distribution is especially hard to fulfil when primary data is used and the sample size is limited. Additionally, covariance-based approaches rely on larger samples, since the sample size needs to be five time higher than the total number of indicators used in the model. Many studies using the covariance-based approach violate this requirement (BURNS et al. 1993:615). These criteria are not valid for the partial least square approach, which is more likely to be used with small samples. A general issue that should be kept in mind is the constraint that only linear dependencies can be modelled, or techniques need to be used simulating non-linear modelling, such as using logarithmised or exponentialised input data (BROSIUS 2008:536; KLINE 2005:23; MARSH et al. 2006:256). Another limitation for the application of SEM is the necessary scale level. The covariance-based approach relies on at least interval scaled data due to the use of the maximum-likelihood estimator, while the variance-based PLS approach is also capable of using ordinal scaled data or dummy variables due to its reliance on the least-square method.

Another critical point for the application and design of structural equation models is the choice between formative and reflective constructs. The majority of authors tend to use formative constructs (FASSOT and EGGERT 2005). This can lead to unnecessary exclusions of indicators, since different quality criteria need to be used for the constructs. In reflective constructs, the observed indicators are influenced by the latent construct. Therefore, all the observed indicators change if there is a change in the latent construct, and the elimination of single indicators is not a methodological problem for the latent construct. This type of construct is mostly used in marketing or psychology to represent, for example, attitudes of people which cause specific actions, i.e. the attitude towards a specific product and the resulting process of buying or recommending to other people. In formative constructs, the latent variable is determined by the observed indicators and measured as a weighted index of its indicators. Therefore, the choice of indicators for the specific latent variable should be carefully founded, and the deletion of one indicator can change the whole meaning of the construct.

5.2 Usage of remote sensing and GIS for vulnerability assessment

Supplementary to the survey data and the focus group data, remote sensing pictures were used. The study used IKONOS satellite data acquired before the tsunami in 2003, after the tsunami in January 2005, as well as an additional set taken in 2008. In order to derive location information, pan-sharpened data with a resolution of one metre were used (ROEMER et al. 2010). The questionnaires of the household and company survey contained various information for locating the study cases interviewed in a Geographical Information System. Besides the UTM coordinates taken with GPS devices, information about the building was gathered to support the exact localisation of the interview object. Roof colour, roof type and building size helped to identify the exact coordination and location of the interviewed household or company in combination with the IKONOS Data in the GIS. In a further step, polygon shapes were created from these localised objects. These polygon shapes were extracted using an automatic method with an object-oriented image analysis containing rule sets in order to classify the image objects automatically. This rule set was developed on the basis of the 2003 IKONOS set and was transferred to the scenes from 2005 and 2008. In order to achieve a better solution, the image object rule set was combined with a digital elevation model of the ground surface derived from MFC-Data with a resolution of 15 cm gathered in a flight campaign in 2008 (STERR et al. 2009). This rule set was developed in a diploma thesis at the University of Munich and applied to the study regions (TEGTMEIER 2010; cp. Figure 23). For this automatic building extraction, an overall accuracy of 91.8 % could be achieved, showing that in less than 10 % of the cases, buildings were not recognised or non-existent buildings were found by the algorithm. Comparing the automatically generated results with manually digitised reference shapes in a 100x100 m² grid shows that the algorithm faces slight problems with the separation of buildings, since in both communities, the number of buildings was underestimated - by 10 % in Ban Nam Khem and by nearly 20 % in Khao Lak. Despite this, the size of buildings in Khao Lak was matched almost exactly, while in Ban Nam Khem, with its higher building density and smaller buildings, the algorithm merged several buildings. This merging of buildings leads to an overestimation of the built-up area by 30 % (cp. Table 12).

Table 12: Accuracy of automatic building classification (Own calculation)

	Ban Nam Khem				Khao Lak			
	Reference		Automatic		Reference		Automatic	
	Mean	Std. deviation	Mean	Std. deviation	Mean	Std. deviation	Mean	Std. deviation
Number of buildings	5.0	5.1	4.5	3.8	4.1	4.4	3.3	3.2
Size of buildings	1034.7	1109.5	1364.9	1387.6	1102.6	1067.9	1103.1	1185.7

The resulting shape of the automatic building extraction procedure could be extended by the building usage information derived from the field mapping. Since this represents the situation in 2008, it had to be assumed that non-destroyed buildings showed the same usage before the tsunami. For destroyed buildings, usage was assigned corresponding to the average building size derived from the household or company survey.

Figure 23: IKONOS Scenes from Khao Lak 2008 demonstrating the outcome of automatic building extraction (TEGTMEIER 2010)

This polygon shape was used as the operational basis for deriving several location factors discussed in chapter 3.2.2. To derive this information, this shape was combined with the digital surface model from the data gathered with the multifunctional camera. This digital surface model has a resolution of around 15 cm. In the GIS, the height information for every surveyed building in metres above mean sea level was extracted. The second indicator forming the location factor for households and companies is the distance from coastline (RUANGRASSAMEE et al. 2006; REESE 2003:199). To derive this shape, the distance in metres from the nearest point on the coastline was calculated in the GIS. As a reference coastline shapes from 2003 were used, which were provided by the non-governmental research organization 'global change **S**ystem for **A**nalysis, **R**esearch and **T**raining (START). The combination of survey data and remote sensing information was also used to derive several items of exposure-related information for the entire group of households and companies in the study regions as a basis for the subsequent investigation. Due to the shortage of official data, this method had to be used to gather information on the micro and meso-scale for households and companies, and on the economic system on the community level (cp. Chapter 4.1).

Statistical model-based prediction of population distribution has been used for a long time to check the reliability of census data or to compensate for shortcomings of census data. WU et al. (2005) reviewed the different methods that used statistical modelling techniques in order to predict the number and the distribution of population in settlement areas. They divided five groups of approaches based on the relationships between population and (1) urban areas, (2) land use, (3) dwelling units, (4) image pixel characteristics, and (5) other physical or socio-economic characteristics. For the assessment of the number and distribution of population on the lowest scale, i.e. the household/building scale, the usage of the relation between dwelling units, pixel characteristics and other socio-economic characteristics is the only suitable method. Linkages between urban areas or land use lead to rougher classifications and distributions which are not useful for the micro-scale vulnerability assessment. The top-down calculation method is widely used, in which census data available on a high scale is combined with census information about the total building size used for housing purposes. Combining this data, the average number of persons per square meter is calculated and applied to each building in the study region. TAUBENBÖCK et al. (2008:82) applied this method to the city in Padang for their assessment of vulnerability in the context of vulnerability estimation; as a result of the application, they reported 'plausibility' for their calculations. This method combines the population information with the number of buildings.

In the TRAIT Project, a top-down calculation for the distribution of population for day and night times was also applied to the complete region - by TIFFERT in a diploma thesis (TIFFERT 2010), covering the area from Khao Lak up to Ban Nam Khem in the north. The dasymetric mapping approach used is based on a land-use classification combined with ancillary data from the household survey (WU et al. 2005). The method uses an algorithm that disaggregates/distributes the total population of the areas classified as settlement areas according to the proportional size of this area in relation to the total area. TIFFERT extended this approach by weighting the settlement areas according to road networks within the area, assuming that a longer road network goes hand in hand with a more dense population. For the other land use classes, the distribution of persons was carried out by weighting the size of the area with the percentage of people working on this kind of classified area derived on the basis of the household survey (TIFFERT 2010:33). For the settlement areas of Ban Nam Khem, the algorithm estimated a total population of 5,063 persons, and 4,613 persons during the day time, by distributing the total number of persons in the area over the different settlements in this area (cp. Map 3).

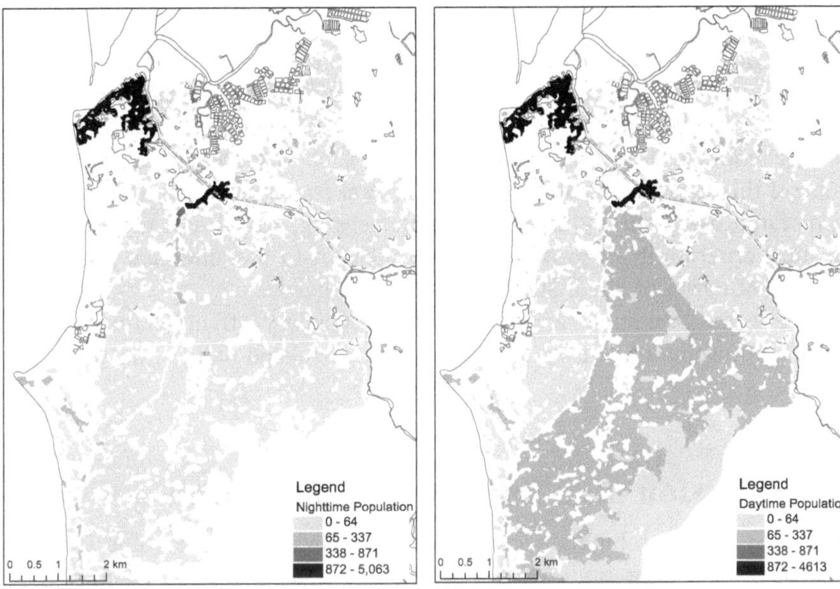

Map 3: *Night and day time residential population 2009 (Author's design, based on Tiffert 2010)*

The method provides a good overview of the situation in a relatively large area when detailed building information is not available or cannot be derived due to budget or time constraints. Since the method depends on the land use classification, it is crucial that the land use is classified correctly. TIFFERT (2010:IX) refers to a Kappa-coefficient of 0.8 for the accuracy assessment, which means that 80 % of the pixels are classified correctly. Besides this, the class 'urban' contains all kinds of buildings from which only hotels have been excluded for extra investigations. Business areas, production halls, temples and other buildings are included in the 'urban' class, and therefore lead to an overestimation of the total number of inhabitants in this area. Additionally, the basis for the classification are the official statistics which cover the 'de jure' size of the population. Since a huge number of people moved away from Ban Nam Khem after the tsunami without changing their registration, an overestimation in some parts of the area and an underestimation in other parts of the area seems more likely. Therefore, for the detailed assessment of exposure in the core communities of Patong Beach, Khao Lak and Ban Nam Khem, the regression-based approach will be used, while for the larger scale, the dasymetric mapping seems accurate.

In a further paper concerning this topic, TAUBENBÖCK et al. (2009) applied another method using physical and socio-economic characteristics of buildings to derive homogeneous settlement areas labelled as 'slums', 'low class', 'medium class' and 'high class', showing similar structural characteristics concerning the building height, density and building size. Using data from their own survey, socio-economic characteristics were linked to each of these classes. In the evaluation of this method, the authors report several constraints arising from empirical research using survey data and remote sensing, but do not give an overall statistical criterion that can work as a goodness-of-fit measurement for their approach (TAUBENBÖCK et al. 2009).

Due to the limitations arising from the secondary official statistics, a bottom-up approach will be used to derive exposure-related indices for the affected regions.

5.3 Discussion of methodological approach

In order to achieve a highly accurate and most comprehensive vulnerability assessment for the four study communities aiming at a generic vulnerability assessment, a multi-method approach was chosen. The evaluation is intended to be based on quantitative and qualitative interview material. Due to the frequent constraints arising from the time-consuming and expensive method of surveys, the aim is to combine this method with remote sensing pictures which are less expensive and available for wider areas. Due to the unexpected occurrence of the disaster, survey data from the time before the event are not available, and only ex-post evaluations of the situation before the onset of the disaster are possible. In the survey conducted in early 2009, attempts were made to compensate for this by asking the interviewees for their pre-tsunami situation in comparison to the situation at the time of the interview. This procedure has the disadvantage that the results might be biased by the fact that an economic situation five years before must be evaluated by the interviewee, which might lead to inaccuracy. Additionally, biased answers caused by over- or underestimating the effects might result from personal interviews (cp. Chapter 4.4). Through the cross-validation with the results from the analysis of the remote sensing pictures and the focus-group discussion results, this bias can be reduced.

Furthermore, this mix of methods allows the coverage of the micro-information about households and companies to be extended and enables the interpolation of this data to non-surveyed households in order to investigate their vulnerability perspective of economic exposure. Following the vulnerability strand of physical vulnerability, this dataset allows an interpolation of the total effects of the hazard in order to estimate the direct damage on the micro-level. This estimation procedure fills a gap in the data, since no official information about micro-level tsunami damage is available, and allows the adaptation of the estimation procedure for further tsunami scenarios. Despite the approaches used in the risk analysis, the subsequent investigation will not be focused solely on the occurring damage, but will rather entail a holistic assessment through the development of a vulnerability model based on structural equation modelling, integrating different elements causing tsunami impacts and integrating the coping capacity of households or companies. The outcomes of these structural equation models for the interrelations among the perspectives of vulnerability is triangulated and proofed with the qualitative results from focus group discussions. In a final step, the two strands of regression and remote sensing-based exposure analysis and structural equation modelling of the vulnerability state are combined on the meso-level for the current situation in the study communities (cp. Figure 24).

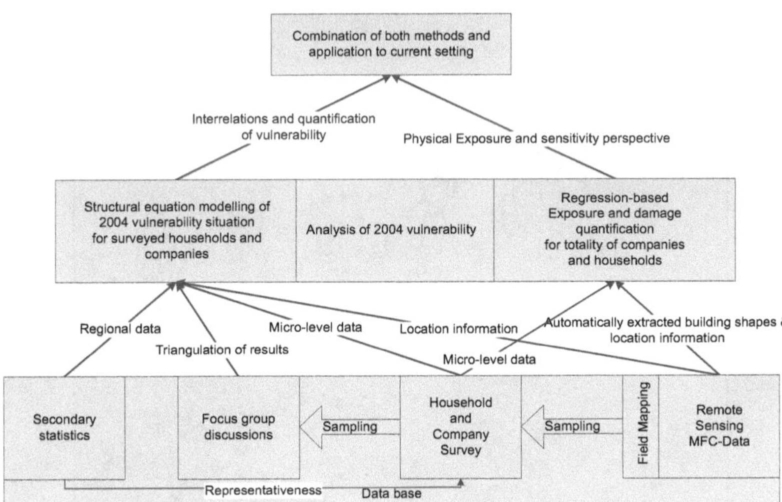

Figure 24: Application of methods and data (Provided by author)

6 Characterisation of the study area and the effects of the 2004 tsunami

6.1 Economic situation in southern Thailand

The tsunami affected the six provinces of Ranong, Phang-Nga, Phuket, Krabi, Trang and Satun in the south of Thailand along the Andaman coastline (cp. Map 4).

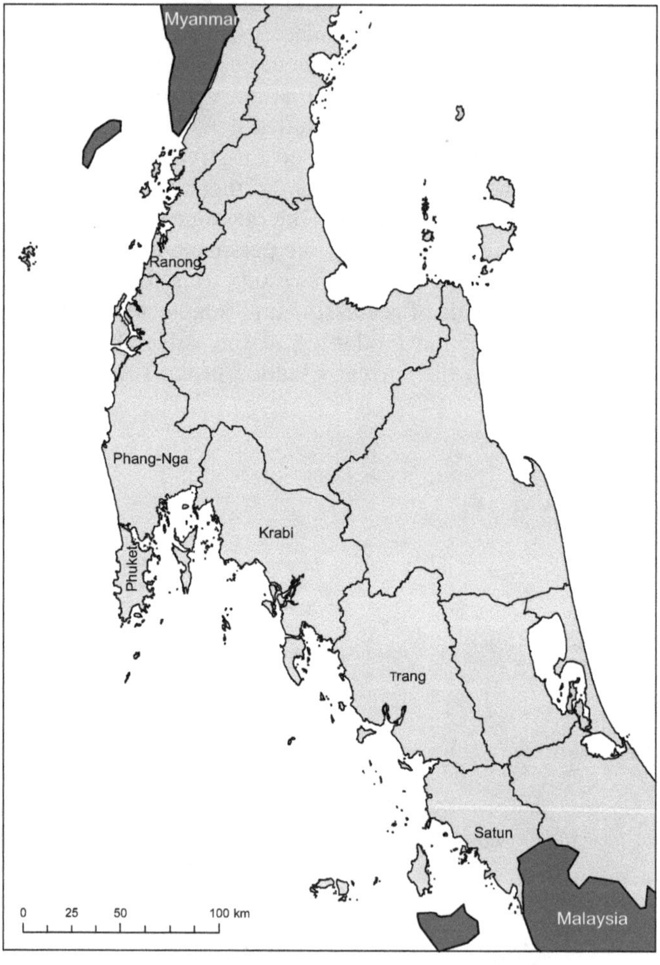

Map 4: Affected provinces on the west coast of Thailand (Provided by author)

Together, the six provinces accounted for nearly 3 % of the total gross domestic product of the Kingdom of Thailand in 2008. The economic situation in these six provinces is highly uneven. While the Phuket Province has one of the highest income levels of all provinces in Thailand, with close to 5,000 € per Person in 2009, all the other tsunami-affected provinces show a significantly lower income than that of the kingdom as a whole. The most southern provinces of Trang and Satun in particular have a GPP which is one-third lower than the GDP. The Phuket Province has a long tradition as an economic hub for the southern region. With the commencement of tin mining on the Malay Peninsula in the 1850s, the island developed into a major production centre and also a trading point for tin (DIXON 1999:51). The tin mining industry was a major income source for some parts of the southern provinces until the 1990s, when the last concessions were phased out (PHAPHAVASIT et al. 2009). Nowadays, no significant contribution of the 'mining and quarrying' sector in the six provinces can be found (cp. Table 13).

Table 13: Economic key facts for the six southern provinces (Author's own design, based on Own calculations; NESDB 2009)

	Phuket	Phang-Nga	Krabi	Ranong	Trang	Satun	Thailand
Population 2003 in 1,000	267.6	248.2	360.0	172.9	630.9	265.9	63,982.1
Population 2008 in 1,000	293.1	262.5	386.4	185.4	671.9	284.3	66,481.7
GPP per capita 2003 (in €)	3,727	1,630	1,735	1,442	1,333	1,457	2,008
GPP per capita 2008 (in €)	4,746	2,596	2,536	2,029	2,069	2,101	2,898
Population per km²	498	57	80	49	123	109	122
GPP according to sectors 2008 as share of the total GPP/GDP							
Agriculture, hunting a. forestry	3.7%	46.2%	46.3%	21.5%	40.7%	30.4%	10.4%
Fishing	4.3%	8.8%	3.6%	25.8%	8.8%	20.7%	1.1%
Mining a. quarrying	0.0%	1.2%	0.4%	0.2%	0.1%	0.5%	3.5%
Manufacturing	4.6%	6.4%	10.9%	7.0%	14.3%	12.6%	35.9%
Electricity, gas a. water supply	2.8%	1.4%	1.9%	2.1%	1.1%	1.3%	2.9%
Construction	5.2%	2.1%	1.9%	3.1%	1.8%	2.7%	2.8%
Wholesale a. retail trade	10.5%	10.9%	8.5%	12.1%	12.6%	9.6%	13.5%
Hotels a. restaurants	38.9%	2.0%	5.1%	1.9%	0.6%	0.5%	4.9%
Transport, storage a. communications	14.1%	3.2%	5.9%	4.7%	3.3%	2.9%	7.0%
Financial intermediation	4.3%	2.6%	2.5%	2.9%	2.4%	1.5%	3.9%
Real estate, renting a. business activities	2.8%	2.2%	2.4%	3.1%	2.2%	2.1%	2.3%
Public administration a. defence	2.3%	3.9%	2.7%	4.6%	2.8%	3.8%	4.3%
Education	2.5%	4.8%	4.9%	5.1%	5.6%	7.3%	4.1%
Health a. social work	1.7%	2.5%	1.5%	2.9%	2.4%	1.9%	1.9%
Other community, social a. personal services activities	2.1%	1.6%	1.7%	2.9%	1.3%	1.9%	1.4%
Private households with employed persons	0.1%	0.1%	0.0%	0.1%	0.0%	0.0%	0.1%

Another business sector in the southern provinces which is still successful is the agricultural sector, building on a long tradition. This sector contributed between a quarter and half of the provincial GPP in 2009, except for the Phuket Province, where its importance has been gradually declining in the past three decades. In total, the southern provinces account for one-third of the total agricultural production of the country (NESDB 2009). Agriculture is one of Thailand's central business sectors and is, in some areas, highly competitive and export-oriented. The country is the world's leading exporter of rice, pineapple, natural rubber, prawns and processed chicken meat (FALVEY 2000:2pp; VORLAUFER 2009:152pp). The most important export markets are Europe and East Asia. Except for the export-orientated agribusinesses, the agricultural sector in southern Thailand is still made up of small family businesses with not more than four hectares of land used for a variety of crops such as rubber, oil palm, coconuts and other tropical fruits. The rubber sector in particular led to a better performance of the southern agricultural sector, and Thailand is one of the largest rubber exporters worldwide. 90 % of these crops are planted in small family-run businesses (VORLAUFER 2009:134pp). In Satun, Trang, Phuket and Phang-Nga, more than 50 % of the total planting area is used for rubber (FAO 2005a:9). Another important business sector along the Andaman coast is the fishing and aquaculture sector. While on the national level, this sector only accounts for a marginal share of the total GDP, in the southern provinces, this share contributes up to 8 %. Nearly 70 % of the country's total fishing production takes place in the southern provinces of Thailand (NESDB 2009). In 2005, the country exported fish worth more than three billion €, and is the third biggest exporter of fish (FAO 2006). Similar to the agriculture sector, the fishing sector in Thailand is split into two groups of producers. Small-scale fishing accounts for around 17 % of the total production measured in tons, whereas the value of this fishing represents a quarter of the total value addition (PANJARAT 2008:39). This group of fishers faces challenges from overfishing, decreasing prices and the dependency on middlemen to sell their catch. Additionally, land use conflicts arise with the booming tourist industry about the usage of beach and coastal areas (VORLAUFER 2009:163). The 'large-scale fishery', on the other hand, is highly competitive and has a much stronger influence on the local income with the upstream and downstream connections, i.e. to ice factories, fish-meal factories, canning, fish-sauce and other processing units. This fishing industry operates fisher boats for coastal fishing with a crew of 10-15 persons or off-shore trawlers. The aquaculture industry is mainly export-oriented and accounts for one-third of the total production of the fishing sector.

These industries of agriculture, fishing and mining made Phuket Island one of the richest parts of Thailand by the 1910s, and allowed the province to expand its infrastructure very quickly (KRACK 2009). In order to reduce the nationwide disparities in income distribution and economic growth, several national plans have been initiated by the NESDB (DIXON 1999:223). The special target was the decentralisation away from the primary city of Bangkok in order to strengthen the remote provinces of the country (NESDB 1992:121). The plans started promoting regional centres, first on the eastern seaboard around the province of Rayong in the early 1980s, and then, in the sixth plan (1987-1991), in the upper southern region (Surat Thani, Phuket, Krabi and Phang-Nga) (NESDB 1987). A major key for this regional development was the booming tourism industry, the potential of which was realised early during the 1970s as a tool for economic growth. In 1977, the government authorities, especially the TAT and the NESDB, started to promote the country for international tourism. In 1981, the tourism sector overtook the agricultural sector in the country and became the most important industry sector. The southern region benefited from this boom very early due to the rich natural heritage, and especially through the upgrading of the small domestic airport on Phuket Island to an international airport in the 1990s (KRACK 2009). The tourism sector is not clearly recognisable and is not stated in national accounts or in the TSIC. Instead, it comprises different sectors such as the hotel and restaurant sector, retail shopping and recreational facilities. Of these, hotels and restaurants are probably the most important segments of the tourist industry (UNITED NATIONS 1991:10). FREYER (2009:125pp) suggests a definition of the tourism sector according to the demand structure of the business sectors. Tourism in the narrow sense combines all businesses producing services directly consumed by domestic or foreign tourists and business travellers. He considers the hotel and restaurant sector as belonging to the tourism sector in the narrow sense. Furthermore, he distinguishes between sectors specialised in tourism, such as souvenir shops or services, and the sectors depending on tourism, such as retail trade or recreational services.

A strong influence of the tourism sector can be found in the directly linked sector 'hotels and restaurants', which contributes more than one-third of the GPP of Phuket. The sector 'transport, storage and communications' also contributes a share to the GPP which is double that of the national level. This sector also comprises businesses for passenger transport and its arrangement, such as travel agencies, tour operators and other local transport companies, for example taxis. While in Phuket, these sectors dominate the whole economy of the province, in Phang-Nga and Krabi, only regional hotspots such as Khao Lak or the Phi-Phi Islands are involved in the tourism business. The average tourist spends nearly 100 € per day during his/her stay in Thailand (TAT 2009c). The total revenue spreads out to several business sectors, from tourism-related industries in the narrow sense to bordering sectors from the regional transportation sector or retail shops. According to the TAT, in the Phuket Province in 2007, less than half of the total revenue generated from tourism was spent in the hotel and restaurant sector. Another 16 % was spent in the retail sector and 12 % in the entertainment sector. For transportation purposes within the country, tourists spent one-fifth of their total budget (TAT 2007b). The inflow of spending power accounted for more than 6 % of Thailand's GDP in 2007 (cp. Table 14).

Table 14: Tourism revenue in million € (Own calculation based on TAT 2009c; NESDB 2009)

	2003	2004	2005	2006	2007	2007 in % of GPP/GDP	2007 tourism revenue per capita in €
Krabi	376.4	406.8	144.1	427.9	556.2	56.9 %	1461.4
Phang-Nga	137.9	205.7	50.3	72.1	87.7	13.4 %	338.1
Phuket	1,578.8	1,803.4	549.3	1,675.3	2,119.5	154.1 %	7330.4
Ranong	21.5	20.8	20.7	22.2	24.9	6.8 %	136.1
Satun	19.4	20.6	27.7	30.7	40.4	6.8 %	144.2
Trang	52.3	67.7	44.7	58.1	68.8	5.0 %	103.7
Thailand	6,664.8	8,090.8	7,160.2	10,412.9	12,319.6	6.5 %	186.5

In the southern provinces, especially Krabi, Phang-Nga and Phuket, the economy is much more dependent on tourism than the other tsunami-struck provinces, and the revenues from tourism range from 13 % of the GPP in Phang-Nga to 150 % in Phuket (TAT 2007a; TAT 2007b; NESDB 2009; own calculations). The reason for these large numbers is that the Bangkok Metropolitan Region can be seen as the heart and the centre of the Thai economy (SCHILLER 2006:3). The majority of economic activities are concentrated here and the majority of national headquarters are located in the region. Since revenues generated in the southern provinces are counted at the location of the headquarters, the GPP shows lower values than are actually achieved. Parallel to this, the tourist revenue per person exceeds the economic performance per capita in Phuket by more than 50 %.

The export-oriented businesses in the fishing sector and the tourism sector, which generate income flows to the southern provinces, result in a significantly higher income than in the rest of the country. Phuket, Krabi and Phang-Nga, which are dominated by fishing and tourism businesses, benefit from these income sources, which also results in a poverty incidence that ranges between

0.2 % in Phuket and 7 % in Krabi, which is still nearly half the figure for the total kingdom (cp. Table 15).

Table 15: Household income and poverty rate in the southern provinces in 2004 and 2007 (Author's own design, based on UNDP 2007:117; UNDP 2010:151)

	Monthly household income 2004 in €	Monthly household income 2007 in €	Poverty incidence 2004	Poverty incidence 2007
Krabi	308	433	6.8%	2.7%
Phang-Nga	328	511	0.6%	--
Phuket	488	577	0.2%	--
Ranong	278	497	5.4%	2.0%
Satun	231	398	6.6%	2.3%
Trang	327	544	2.2%	1.9%
Thailand	289	429	11.3%	8.5%

The higher income in the region does not correlate with more secure income sources for the households. While in the whole country, around one-third of the total workforce is employed informally, that means without an official contract, the share in the southern provinces is much higher. Here, 60 % of the total workforce is employed informally (NSO 2008b). Additionally, the percentage of people working as self-employed individuals or as employers is slightly higher than in the rest of the country. Both facts may result from the low share of industry and the strong focus on service businesses and agriculture and fishing, which are strongly characterised by informal employment. Furthermore, in the leading sectors of fishing and agriculture, the percentage of employers among the total workforce is around 1.5 % higher than in the country as a whole. In the Global Entrepreneurship Monitor 2006, VIRASA et al. (2006:9) found that the population in southern Thailand has a higher acceptance of entrepreneurial activities than in the rest of the country and is more optimistic about recognising the chances for their own business.

6.2 Tsunami 2004

Tsunamis can be triggered by earthquakes under the sea, landslides, volcanic eruptions or by meteoroids which generate a sudden, vertical displacement of a column of water. Typical waves caused by wind have a wave length of around

150 m and a period of ten seconds. The wave length of tsunami waves is more than 100 km with a period of one hour, and they travel at a speed of around 700 km and encompass the whole water column from the floor to the surface of the ocean. The majority of tsunamis occur along the seismic hotspots in the Pacific Ocean with around 1,300 known events. Around 300 Tsunamis have also been reported in the Mediterranean Sea in the last 2000 years (cp. Map 5).

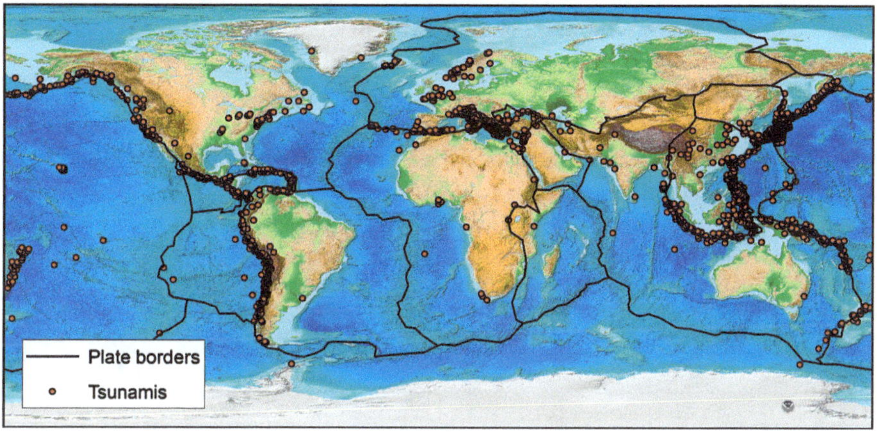

Map 5: Tsunamis since 2000 B.C. (Author's design, based on AMANTE and EAKINS 2009, NOAA 2010)

The majority of tsunamis occur in the Pacific Ocean, and especially along the Japanese coastline, and the term tsunami is indeed of Japanese origin and means harbour wave. The tsunami of 26 December 2004 that devastated large parts of the study area was caused by an earthquake at 8 o'clock in the morning with a magnitude of 9.3. The epicentre was located at the northern tip of Sumatra Island, around 550 km from the Phuket and Phang-Nga Provinces in southern Thailand. Magnitudes like this have been measured for very few earthquakes in the last 100 years. This earthquake caused a tsunami with run-up heights on the coastline of Sumatra of 35 m. Patong Beach in the province of Phuket was hit by waves of up to 8.5 m, while in Khao Lak and Ban Nam Khem, the waves reached 12 m. The waves reached Thailand around 100 minutes after the earthquake that triggered the tsunami. Besides Thailand and Indonesia, India, Sri Lanka, Kenya, Somalia, South Africa and the Maldives were also hit by the waves. The official total numbers of victims reached 250,000 in all affected countries.

6.3 Effects of the tsunami on the economy of Thailand

In total, 230,000 people were killed by the Indian Ocean tsunami of 2004, most of them in Indonesia, Sri Lanka and India (EM-DAT 2010). According to official numbers published by the Department of Disaster Prevention and Mitigation of Thailand (DDPM 2005b), more than 8,000 people lost their lives and more than 8,000 people were injured. The majority of the victims officially counted - around half - were foreigners, mainly tourists from Europe and Australia.

In terms of the financial impact of the tsunami, Thailand was the second most affected country, with direct damages and indirectly induced losses of around 1.8 billion €. Indonesia, where the tsunami was triggered, suffered losses of more than twice this figure (cp. Table 16).

Table 16: Financial impact on tsunami-affected countries (Author's own design, based on UNITED NATIONS *and* WORLD BANK *2005:18)*

	Direct effects	Indirect effects	Total
Indonesia	2.5	1.3	3.7
Thailand	0.4	1.4	1.8
Sri Lanka	0.8	0.3	1.1
India	0.5	0.4	0.9
Maldives	0.3	0.1	0.4
Damage and Losses (Billion €)			

Taking a deeper look into the figures, it can be seen that the direct effects caused by the tsunami, i.e. the damages caused by the impact, only accounted for 0.4 billion €, while the indirect effects, i.e. losses of income or loss of sales, accounted for 1.4 billion €. According to these numbers, the direct impact of the tsunami caused one-third of the total damage, while the indirect effects accounted for two-thirds of the total damage. As a consequence, the growth of the GDP was slowed (UNITED NATIONS and WORLD BANK 2005:10). This index is often used to measure the vulnerability and the recovery potential of sub-national regions (NAUDÉ et al. 2008:1; BRIGUGLIO 2003). In 2004, the NESDB expected a growth of 6 % for 2005. Deducing the loss of added value and the damage caused by the natural hazard, a reduction of 1.5 % for the economic growth of the country could be observed (ADPC 2006:22). In terms of indirect effects of the tsunami, Thailand was the most affected country. The reason was the clustering of businesses along the coastline, meaning that businesses suffered nearly

80 % of the total direct damages. The worst affected sectors were tourism, fisheries and agriculture. In contrast, households suffered only around 5 % of the damage. Of this, 80 % had to be invested for reconstruction or repairing residential buildings, while the rest was needed for restoring personal belongings and household goods. The public sector, i.e. infrastructure, public buildings etc., accounted for around 12 % of the total damage. The total amount of damage to agriculture and business property accounted for more than 7 % of the total GPP of these six affected provinces (NESDB 2009; own calculations). The area hardest hit was the province of Phang-Nga, where the direct damages accounted for more than one-third of the GPP, and business property was destroyed worth approximately 60 % of the economic performance achieved in 2004. Looking at the different business sectors, fisheries suffered from tsunami impacts causing damage to business property of more than 8 % of the GPP generated in 2004 in the six tsunami-affected provinces in this sector. A second business sector which is mostly located near the seashore accounted for even higher losses. In the hotel and restaurant sector, the direct damages rose to nearly half of the sector's contribution to the 2004 GPP. Here, the province of Phang-Nga suffered the most, with losses equivalent to more than seven times the annual GPP of 2004, while in the other provinces, the share ranged from 16 % in Phuket to 110 % in Krabi.

The indirect effects resulting from the tsunami impact showed a contrary picture. In the long term, secondary indirect effects resulted from the reduced spending power of the local population as well as the reduced number of visitors in the tourism sector, who tended to avoid the affected areas (SMITH and WARD 1998:51p). The most obvious indicator for the indirect effect is the GPP measuring the economic performance of the province. While it is assumed that the GDP of Thailand was reduced by 1.5 %, a more diverse picture can be seen in each of the six affected provinces (cp. Table 17).

Table 17: Annual change in GPP/GDP in relation to the previous year in the tsunami-affected provinces (Own calculations based on NESDB 2009)

	2004	2005	2006	2007
Krabi	3.7%	-0.6%	12.1%	2.2%
Phang-Nga	10.1%	1.8%	2.8%	3.6%
Phuket	17.0%	-10.0%	9.7%	4.8%
Ranong	16.6%	2.4%	4.9%	3.1%
Satun	9.2%	6.1%	7.9%	1.1%
Trang	5.1%	7.4%	3.6%	2.5%
Thailand	6.3%	4.6%	5.2%	4.9%

The economic growth in most of the provinces was only dampened. Even in the Phang-Nga Province, which was directly affected the most, the GPP still showed a slight rise. On the other hand, in the Phuket Province, where the direct losses were equivalent to no more than 16 % of the GPP, the highest reduction of economic performance with a shrinking GPP of 10 % could be observed. Still, it can be seen that the economy of the provinces recovered very quickly from this external shock and achieved the same level as before the tsunami by the end of 2007.

In the aftermath of the tsunami of 2004, the revenue from tourism dropped dramatically in the southern provinces (cp. Table 14). The highly tourism-dependent provinces of Phuket, Phang-Nga and Krabi in particular recorded decreases of more than 60 % compared to the previous year. In 2005, the number of visitors to Phuket and Phang-Nga dropped to 85 % of the expected number and resulted in a calculated loss of revenue of 47 % or 509 million € in Phuket and 21 % or 14.2 million € in Phang-Nga for the year 2005 (NESDB 2010; own calculations). The consequences for the Thai tourism industry over the whole kingdom were not so severe. The revenue increased by 4.5 % mainly by shifting the tourist flows away from the south towards the eastern seaboard and the northern provinces (CHAIPAN and CHAIPAN 2006:5). The revenues across the whole nation were able to recover within one year, while the southern provinces needed between one year in Krabi and two years in Phuket. The tourism revenue in Phang-Nga has still not reached the level of 2004, which is also due to the ongoing political crisis which started in 2006 (MCDOWALL and WANG 2009:353). The TAT projected the number of tourists for the six tsunami-affected provinces before the event to be seven million in 2005 and 7.5 million in 2006. The actual numbers for 2005 and 2006 fell below this projection by 5.5

million and 3.5 million respectively. This caused a loss of tourism revenue in comparison to the projected numbers of 1.5 billion € in 2005 and 900 million € in 2006 (own calculation). These numbers of indirect losses result from the consequences of the devastation caused by the tsunami destroying a huge number of tourism facilities in the southern provinces. In all six provinces, more than 300 hotels and resorts were affected by the waves, causing rebuilding or renovation costs to buildings and interiors of more than 0.5 billion €. The major share of this occurred in the provinces of Phang-Nga, where nearly half, and Phuket, where more than one-third of the damage to hotels and resorts was suffered. Looking at the bordering sectors, especially the retail and restaurant sectors targeting tourists, over 4,900 company facilities were destroyed or damaged. The damages, including costs for rebuilding or restoring buildings and interiors, totalled more than 20 million €. While in Phang-Nga, the number of affected companies was below 1,000, on Phuket Island, nearly 3,000 companies were physically affected by the wave impact. Therefore, the average damage per affected company in Phang-Nga reached nearly 130,000 €, while on Phuket Island, the impact per company was only 30,000 €. As a result, the TAT counted a number of hotels for 2005 in the Phang-Nga Province which was only one-third of the number of establishments prior to the natural disaster, and the number of rooms was also reduced to one-third of the previous figure. In Phuket, the hotels seemed to be only slightly affected. Here, the number of hotels in 2005 compared to the previous year was reduced by one-third and the number of rooms by only 6 %.

The second business sector on which the economies in the southern provinces rely is the agricultural and fishing sector. The tsunami waves had a severe impact on the agricultural land located near the coastline and either washed away the crops or deposited salt in the ground. Several plantations with oil palms and coconut trees were uprooted and needed to be replanted. The fishing sector, which is naturally located near the coastline, was also heavily exposed to the tsunami waves. Besides the damage to fishing boats and equipment, the infrastructure of aquaculture was severely damaged. In all six southern provinces together, nearly 2,000 hectares of agricultural land was affected in this way. The most affected area was the province of Phang-Nga. Around 1,000 farmers were affected in their livelihood activities by the tsunami (FAO 2005b). The ADPC (2006:7) estimated the damage to crops at around 6 million €, mainly in plantations, and the cost of the reduction in future production caused by salt water intrusion and the time lapse before replanted trees could reach maturity at two million €. Besides the farming, livestock breeding was also affected. The MOAC reported that nearly 14,000 productive livestock were lost, causing a total damage of 370,000 €. The majority of damage, with 77 %, was located in Phang-

Nga. The production in the fisheries and aquaculture sector was greatly hindered by the destruction of boats, fishing equipment and agricultural equipment. The actual amount of lost fishing and agricultural equipment is difficult to assess, since there is no reliable registration system available. As a consequence, the estimations are based on reports of the fishermen in order to obtain compensation from the authorities. Since compensation rates are low and do not cover the actual repair costs, it is assumed that some fishermen did not report the loss of their assets (cp. Chapter 6.4). According to official numbers of the Department of Fisheries, more than 5,000 boats were damaged or lost, accounting for more than one-third of the total number of officially registered vessels, making up a total damage of 0.25 billion € (FAO 2005a). The amount of damage of fishing equipment, i.e. traps and nets, totalled another nine million € (ADPC 2006b). The aquaculture sub-sector also suffered severe losses to its production facilities. In this sector, the amount of damage reached 22 million €, including the damage to fish cages, shrimp ponds and hatcheries. The hot spot of damage was located in the Phang-Nga Province, where nearly seven million € worth of aquaculture equipment and infrastructure was damaged, representing one-third of the total damage in the six affected provinces of southern Thailand. Setting the total damage in the provinces in relation to their GPP reveals that the provinces of Phuket and Phang-Nga in particular suffered losses of 10 % and 42 % respectively of their annual GPP in 2004. In the other 4 provinces that were also affected by the tsunami, the damage was equivalent to not more than 6 % of their GPP (cp. Table 18).

Table 18: Damage to livelihoods in Euro (Author's own design, based on NIDHIPRABHA 2007:4)

	Fishery	**Livestock**	**Agriculture**	**Business property**	**Total damage in million €**
Phang-Nga	12.4%	0.2%	0.0%	87.4%	156.2
Phuket	8.9%	0.1%	0.0%	91.0%	81.6
Other affected Provinces*	16.9%	0.1%	0.1%	82.8%	68.7
Total	12.5	0.1%	0.0%	87.3%	315.9

* Krabi, Ranong, Trang, Satun

The total impact on the fishing and aquaculture sector totalled nearly 55 million €. The loss of catches in the fisheries sector and production losses in the aquaculture are estimated at around 82 million €. In the end, this came to a total loss of 136.5 million €, showing that the direct effects accounted for 40 % of the total losses, while the indirect effects had a much higher share with 60 %. Following the production chain, the food processing industries were also affected by the impact on the primary sector and the resulting shortage in supply. These downstream industries, i.e. the agro-industry and agro-commerce, recorded losses due to the tsunami of 77 million €, mainly because of the shortage of fish. Besides the shortage of supply, the decline in demand also played a significant role. The fishing communities along the coastline "directed a significant part of their activity to supplying restaurants and resorts with fresh, highly-prized reef fish species and sea food products in general" (NACA 2005:2). This demand was interrupted by the tsunami and the abrupt decrease in tourist numbers (cp. Table 14).

The tsunami also affected a large number of housing areas in the six southern provinces. More than 3,300 houses were destroyed and another 1,500 residential buildings were damaged. The most severely affected area was the Phang-Nga Province, with 2,500 destroyed or damaged buildings totalling more than 10 million €, and the province of Phuket, with 1,000 houses and 4.7 million € worth of damage. While the households in the Phang-Nga Province had an average damage to property of 3,900 €, the households in the Phuket Province suffered from an average damage around 14 % higher (ADPC 2006; cp. Table 19).

Table 19: Damage to the housing sector (Author's own design, based on ADPC 2006:4)

	Number of houses		**Amount of damage (million €)**			
	Destroyed	**Damaged**	**Reconstruction**	**Repair**	**Household Goods**	**Total**
Phang-Nga	1,904	604	6.86	1.06	1.90	9.82
Phuket	742	291	3.13	0.59	0.93	4.67
Other affected Provinces*	656	609	2.03	0.88	0.43	3.34
Total	3,302	1,504	12.02	2.53	3.27	17.83

* Krabi, Ranong, Trang, Satun

Besides the direct impacts, households also suffered from indirect tsunami effects resulting from the loss of income. As seen before, the 'agriculture and fishing' and the 'hotel and restaurant' sectors account for the majority of the economic performance in the southern tsunami-affected provinces. According to the nationwide cost structure published in the input-output table of Thailand 2005 by the NESDB, the 'hotels and restaurants' spent 11 % of their sales volume on wages and salaries, while these expenses accounted for 16 % of the sales volume in the 'agriculture and fishing sector' (NESDB 2006). In the first year after the tsunami, the total volume of wages and salaries in the six provinces dropped by 19 million €, or around one quarter compared to the year 2004. Much more seriously affected were the provinces of Krabi and Phang-Nga, where the total volume of wages and salaries decreased by 38 % and 43 % respectively (NESDB 2006; NESDB 2009; own calculations). The expenses for wages and salaries have still not reached the level of 2004, although more as a consequence of the political instability than of the natural disaster (McDowall and Wang 2009:353). Only the 'hotel and restaurant sector' in the provinces of Satun and Ranong could increase spending on wages and salaries. Traditionally, these provinces have a much higher share of domestic visitors - more than 80 % - while the other southern provinces rely mainly on international tourism and can only attract one-third or less of their visitors from Thailand, who were called upon to compensate for the reduction in the number of international tourists (TAT 2009c; own calculations). In the 'agriculture and fishing' sector, the overall expenses for wages and salaries decreased in the first year after the tsunami by nearly 19 million €. This reduction was mainly caused by the losses in the provinces of Phuket and Krabi, where the wages and salaries dropped by up to 5 %, while in the other provinces, a slight growth was visible. In the Phuket Province, where the fishing sector accounts for more than 60 % of the agricultural and fishing sector, nearly the entire fishing fleet of 1,250 vessels was lost and 70 % of the aquaculture area was destroyed (NESDB 2009; NACA 2005:23). In 2006, the wages and salaries increased compared to 2004 in all provinces except Phang-Nga. Only in Phang-Nga have the wages and salaries still not reached the level of 2004.

The indirect effects resulting from the physical damage to household and company property and the loss of income for the economy were not only limited to the coastal areas and the Amphoes (districts) near the coastline, but also extended to the Amphoes located inland. This occurred particularly through the interrelations within the economic system determined by the sales and supply linkages, the connections to the households determined by work relations (shortage of labour, reduction of income) and the interrupted flows of spending power. Throughout the provinces, the revenue from corporate taxes and the income tax for households dropped by up to 60 % and 30 % respectively. Here, it was not the districts directly affected which suffered the most, but the districts around the provincial capital of Phang-Nga (cp. Map 6). The reason for these disparities in the tax revenue was the postponing of taxes by the government for the households and companies directly affected by the tsunami (cp. Chapter 6.4). Low income or informally employed persons did not benefit from this recovery plan of the government because of the special nature of the Thai tax system. The Thai tax system for households is based on a progressive tax rate determined by personal income. This system consists of five income categories with specific tax rates. Incomes of less than 150,000 THB after deductions and allowances are exempted from paying taxes. For incomes higher than 150,000 THB, the tax rate increases progressively up to 37 % (REVENUE DEPARTMENT 2008). Additionally, in some business sectors, especially the service sector, the majority of workers are employed in an informal way or self-employed (NSO 2008b). These groups are also not included in the taxation system. In the southern provinces, around 20 % of the households do not reach an income above 150,000 THB, so they do not benefit from the tax postponing plans of the government.

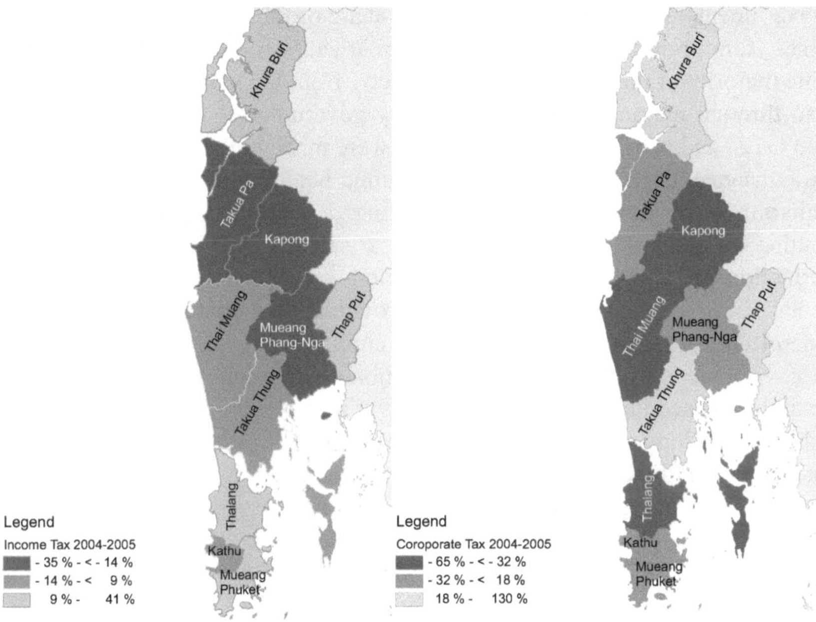

Map 6: *Change in personal income and corporate tax 2004-2005 in % (Own calculation based on NSO 2007a; NSO 2007c)*

6.4 Disaster management in the aftermath of the tsunami

6.4.1 Emergency and recovery support

In the aftermath of the tsunami, several foreign governments offered financial assistance and debt relief for Thailand. These offers were rejected because of the fear of a downturn of the credit ratings for the country and, as a result, an increase of interest rates which the country has to pay for borrowing money on the world market. The country also wished to avoid sending negative signals to foreign investors that the country is dependent on external financial help to solve its problems (KORFF 2009:132). Therefore, only technical assistance from foreign governments was accepted. The biggest share of the donations for technical assistance came from the United States of America, European countries and Japan. The majority of the donations came directly from the national governments, but also through national NGOs financed by governmental funds, such as the German GTZ, and many others. Investments were made in the sectors of humanitarian assistance, environmental projects, public health and infrastructure. The humanitarian sector covered projects providing first aid assistance such as the distribution of clothes, food and other necessary items, and several projects for risk reduction and mitigation were also carried out in this sector. In the environmental sector, projects were supported providing support for the assessment of tsunami damage to mangroves, cleaning coral reefs or providing assistance for an environmentally friendly recovery. Public health and infrastructure projects mainly focused on medical support, providing psychological assistance and rebuilding sanitation and water facilities. The majority of the projects were concentrated in the Phuket and Phang-Nga Provinces (TICA 2006). The total amount of foreign aid reached 110 million €, with another 45 million € provided by companies, mainly foreign firms doing business in Thailand. The total amount of foreign assistance represented one-tenth of the national budget for tsunami relief.

Governmental emergency response started immediately after the event. The Ministry of the Interior set up a centre for assisting the tsunami victims on the same day. The government spent a total of 1.37 billion euro for the emergency assistance, restoration of infrastructure, assistance to companies and restoration of private housing (cp. Table 20).

Table 20: Financial assistance provided by the Thai government (Author's own design, based on TICA 2006)

	Budget in million Euro	Disbursed in million Euro 2006	Disbursed per recipient in Euro
Assistance to affected foreign tourists	1.9	1.9	509
Assistance to Thai victims	13.7	16.4	101
Assistance to fishermen	16.7	11.6	434
Unemployment assistance	4.5	4.5	71
Assistance to SMEs	n/a	4.3	409
Assistance for housing	1.6	1.0	514
Assistance to affected school students	9.4	9.4	271
Assistance to Government officials	8.4	4.6	n/a
Company soft loans	1,250.1	794.4	109,838
Restoration of small scale infrastructure	6.6	5.5	n/a
Restoration of governmental infrastructure	57.4	44.2	n/a
Total	1,370.3	897.6	

For companies in total, nearly 810 million € were made available. The funds were split up in soft loans for big companies with low interest rates of 2 % for the following five years (INTERVIEW KHAO LAK TOURIST ASSOCIATION). These were mainly large hotels near the coastline that suffered heavy losses. More than 7,000 companies participated in this programme in total, with each of the companies receiving on average 110,000 € in loans. A special programme granting financial aid for SMEs was also designed, in which more than 10,000 companies, mostly store vendors, took part (NIDHIPRABHA 2007:20). On average, each of the companies received a total amount of 410 €. According to the Ministry of Industry in Thailand, SMEs in retail have not more than 30 employees and a fixed capital of not more than 1.3 million € (MOI 2003). Another large programme was directly targeted at the fishermen affected by the waves. They received subsidies in total of nearly 17 million €, with each of the 26,600 fisherman receiving around 434 €. The programme supported investments in fishing boats, hatcheries, fishing tools and also tourist boats. This programme did not cover the complete cost of the disaster, since the DDPM refers to a total damage in the fisheries sector of more than 38 million €. NIDHIPRABHA (2007:19) explains this difference between the needs and the offer of financial assistance from the government through the employment effect that justifies the favouritism of bigger companies. Furthermore, taxes and loans for individuals and businesses to the government were reprieved in order to reduce the financial impact on the affected region. The second largest budget point was designated for the restoration of infrastructure. Around 7 million € were invested to cover small-scale infrastructure damage on roads, bridges and piers. By the end of 2006, 5.45 million € had already been spent, covering 90 % of the estimated damage of 6 million €. Another 45 million € were designated to cover the damage to government property, i.e. the Royal Army, hospitals, national parks, schools and colleges, the Royal Thai Police, energy supply and religious institutions, and the additional budget needed to carry out the rehabilitation measures.

For private persons and households, several other supporting projects were designed in order to give short and long-term relief from the consequences of this natural hazard. The largest budget point targeting private persons was the personal assistance for Thai victims. This programme covered compensation for personal casualties and funeral expenses. Tools for livelihood, kitchen utensils, beddings and other furniture or long-term consumable assets, such as TVs, refrigerators etc., were also given to the affected people, mostly by the Royal Thai Army. For the death of a family member, 421 € in compensation were paid to the family, while injured people received between 40 and 100 €. Affected families could also claim assistance in the form of 40 €. Additionally, people who had lost their jobs could claim unemployment support. This was done by more than 62,000 people, with 40 € being paid to each applicant on average, which is the equivalent of a quarter of an average monthly income for one person in southern Thailand (NESDB 2009). Besides the payment of unemployment help, employment projects were started and occupational training was offered to the jobless. An additional fund was created to compensate tourists who were injured by the waves and family members of tourists killed in the disaster. This compensation given to more than 3,600 tourists was nearly five times higher than that provided to the local population. Besides the higher costs for funerals and transport back to the home country, the concerns about the reputation of the country as a tourist destination might have played an important role.

According to official numbers from the DDPM, 4,806 houses in 407 villages were destroyed or damaged by the wave impact (DDPM 2005b). Governmental agencies were asked by more than 4,500 people for assistance in rebuilding their houses. On average, around 500 € were spent for each of these houses. According to official numbers, 3,414 applications were approved and, additionally, 544 persons received 630 € in compensation to rebuild their houses on their own. These projects were mainly carried out by the Royal Thai Army and volunteers. Additionally, several NGOs built houses in the provinces of Ranong and Phang-Nga (cp. Figure 25).

Figure 25: House build by governmental agency in Khao Lak (left) and House build by NGO in Ban Nam Khem (right) (Own Photo)

Unfortunately, no figures for the total number of houses built are available for the projects carried out by the NGOs. Comparing the figures between the actual numbers of houses built or restored and the number damaged or destroyed, some irregularities are obvious (cp. Figure 26). The governmental help focused on the northern provinces of Phang-Nga and Ranong. In these two provinces, more houses were repaired or rebuilt by the government than were counted by the DDPM as having been destroyed or damaged. In the provinces of Krabi and Trang, 25 % and 20 % respectively of the damaged or destroyed houses were restored or rebuilt by the government. In the Phuket Province, the governmental programme only covered 12 % of the actual damage, and in the most southern province of Satun, no compensation was granted (cp. Figure 26).

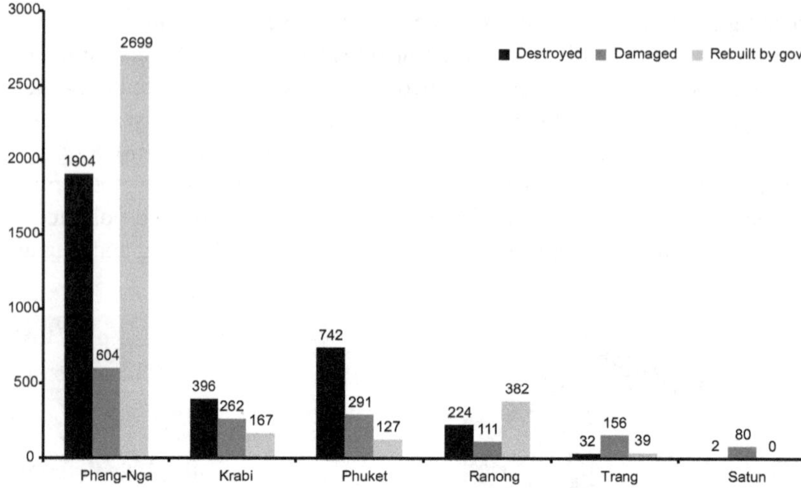

Figure 26: Tsunami impact on housing and rebuilding (Author's design based on DDPM 2005b; RATTANANAKIN 2006:3)

6.4.2 Coordination of support

The recovery efforts undertaken by the national governments, international and domestic NGOs, and foreign governments in the tsunami-affected regions were widely described as the 'largest ever' (JAYASURIYA and MCCAWLEY 2008:4). Because of the strong media coverage with the dramatic pictures of the rapidly occurring natural hazard from these globally know tourist areas, the inflow of financial support was enormous. As well as the chance of a fast and sustainable recovery process, this time period also carries a variety of risks. Despite the rejection of international help by the Thai government, a variety of offers of assistance from governments, multinational donor organisations, companies and individuals came to the country. Therefore, the coordination of this massive amount of help was the most pressing issue in order to guarantee effective delivery and distribution of help. Since the local administration on the municipality level, and even on the provincial level, could not handle the effects of this major natural disaster on their own, the responsibility went to the national authorities to coordinate the relief efforts (KAMOLVEJ 2006:74). Since the governors of the provinces are under the authority of the MOI, this governmental agency in Bangkok holds the responsibilities for coordinating the measures in all the affected provinces. For the distribution and coordination of the relief measures, the government founded the "Southern Disaster Victim Relief Collaboration

Centre" in Phuket. This Centre was put under the control of the MOI and the other ministries involved. The main function was the coordination of assistance to all affected provinces. Furthermore, responsibilities were zoned, with the MOI itself being responsible for the Phuket Province, the Ministry of Natural Resources and Environment for Phang-Nga, the Deputy Minister of the Interior for Krabi, and the Deputy Prime Minister for Ranong. These ministries were in charge of inspecting and solving the problems in their area of concern in cooperation with the governmental organisations in charge (KAMOLVEJ 2006). Additionally, a relief centre at the DDPM in Bangkok was established to coordinate the mobilisation of relief efforts from all sectors of society. It also provided disaster-related information to relatives of disaster victims and a donation centre to receive and coordinate donations. The relief efforts were administrated and conducted by the separate ministries (DDPM 2005a). For example, the MOI was responsible for providing assistance to small-scale business entrepreneurs, the Ministry of Finance provided assistance to large-scale business entrepreneurs, and the Ministry of Agriculture and Cooperatives assisted the fishery sector. Private persons affected by the tsunami were compensated by the Permanent Secretary for the Office of the Prime Ministry, and the Ministry of Labour provided assistance to the unemployed.

In total, 393 organisations could be identified that carried out activities in the region. Two-thirds of them were governmental organisations. The majority of the organisations that were involved in risk management were national organisations or provincial organisations, and only 15 % were from the local level. Using the technique of network analysis, KAMOLVEJ (2006) identified 531 ties between the organisations with main hubs in the MOI and the Royal Thai Navy. These hubs also initiated around one-fifth of the total interactions within the network. This corresponds to the official plan of the guiding role of the Ministry of the Interior, coordinating the efforts, and the Royal Thai Navy, which was responsible for rebuilding and for donations to the local populations. This network analysis showed that despite these central hubs, the network is not very strongly connected, which indicates a lack of coordination and cooperation. Also, the central role of the country's governmental ministries indicates that the network suffered from a lack of information on the real needs in the affected areas, since the coordination took place in the capital Bangkok and also in Phuket. The huge number of organisations involved in the recovery process led to several redundancies within the structures, for example in cases where missing persons were counted twice when their absence was reported to different organisations (KAMOLVEJ 2006:76). After a fast and effective emergency response in the first days and weeks after the disaster, this lack of organisation led to several problems in the disaster management (NIDHIPRABHA 2007:18). Of the total donation budget designated by the government for the recovery, nearly 20 % had been disbursed among the local population and companies after two months. The figure went up to 70 % after one year and then slowed down significantly to reach 80 % after two years. Larger donations from international organisations in particular showed a slow disbursement rate because some of them were more targeted at the long-term building of mitigation capacity. On the other hand, a rapid disbursement of the total available donations could also lead to problems, especially in a natural hazard occurrence affecting a huge area like the tsunami of 2004. Large scale reconstruction programmes have a huge demand for input in the form of material and labour. Therefore, a positive economic development can often be observed after a natural disaster, leading to a shortage in labour and increasing prices for goods that are needed in the reconstruction programme. In the worst case scenario, this can lead to the so-called 'Dutch Disease', where the booming business sector absorbs labour force and material input from other sectors so that other non-booming sectors suffer from rising costs and fall into a downturn. In small countries in particular, where the pool for labour and material is not big enough to compensate for the additional demand, the prices tend to be inelastic and rise. This was also feared for the Thai economy. Thailand's economy was able to absorb the additional demand for building material and labour without cost escalations, in contrast to the situation in Sri Lanka or Indonesia, where the costs increased by between 50 % and 80 % (WEERAKOON et al. 2007:33). The increased demand for construction could be easily compensated for by the country's economy, because the potential of the metropolitan region of Bangkok could be used, and the construction sector in Thailand was in a general depression, which compensated for the increased demand in the south. In the following period, only minor increases of 5 % could be observed for wood and wood products, while steel prices dropped, shrinking overall construction costs as a consequence (NIDHIPRABHA 2007:11).

6.4.3 Evaluation of emergency aid and recovery support

The rehabilitation programmes showed a positive influence on the regional employment. The unemployment rate in the southern provinces before the tsunami was around 1.5 % (BoT 2010)[1]. In the first quarter of 2005, the unemployment rate in the Phuket and Phang-Nga Provinces nearly doubled, since in some areas, the majority of businesses had to close and let their workforce go (cp. Figure 27).

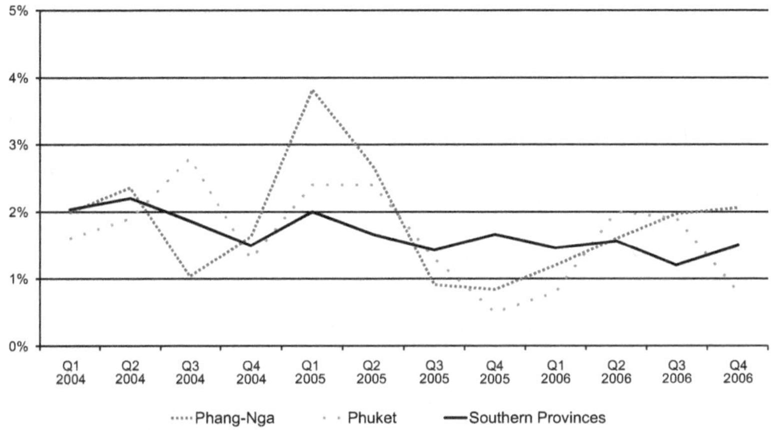

Figure 27: Unemployment rate (Author's design, based on data from BOT 2010; NSO 2007a; NSO 2007c)

In the third quarter of 2005, a significant decrease could be seen, which brought back the unemployment rate to the level before the tsunami, and in the case of the Phuket Province, to an even lower level, which might also partially have been caused by short-term outward migration. A similar picture can be seen when looking at the GPP normalised to the year 2004 (cp. Table 21). In the Phuket and Krabi Provinces in particular, a drastic reduction of the economic performance was evident in 2005, while the growth curve in the other provinces, such as Phang-Nga, Ranong and Satun, Trang, and in the country as a whole was dampened only minimally. One year after the tsunami, the economy in Krabi and Phuket started to grow again, and Krabi at least could catch up with the economic growth of the southern provinces after two years. The factors deter-

1 The official unemployment rate doesn't represent the totality of the unemployed workforce. Informal employed persons are not entitled to claim unemployment support. Since formal employment in some sectors is not more than 50 % these persons are not included in this statistic (NSO 2008b).

mining these developments will be further investigated by looking at the special settings in four communities in the provinces of Phang-Nga and Phuket.

Table 21: Gross provincial product indexed to 2004 (Author's design based on data from NESDB 2009)

	2004	2005	2006	2007	2008
Krabi	100	99	112	114	112
Phuket	100	90	99	104	105
Ranong	100	102	107	111	115
Phang-Nga	100	102	105	108	109
Satun	100	105	110	116	116
Trang	100	107	111	114	113
Thailand	100	105	110	115	118

6.4.4 Reputational risk management

As outlined already, tourism is one of the major income sources of the provinces in southern Thailand. Since tourism is dependent on the perception of the worldwide tourism market about safety and the situation at the destinations, a crucial point of recovery was the mitigation of the concerns of potential tourists about returning to the tsunami-affected areas. In order to assess these efforts, the region will be compared with the Maldives, which were able to return faster to their previous tourism levels after the disaster. This country is comparable to the study region in terms of dependency on tourism, since around one-third of the GDP and 40 % of tax revenues are generated by the tourism sector (MPND 2009). The sector is, like southern Thailand, dependent on the key markets in Western Europe (CARLSEN and HUGHES 2008). Comparable to Thailand, the direct losses totalled 214.7 million € and the indirect losses 132.1 million €, which is equivalent to around 62 % of the country's economic achievement for the year 2004, due to its overall relatively small economy. The government assessed losses of 176 million € in total for the tourism sector alone, which employs around 25,000 people, either directly or including multiplication effects, in the country of 314,000 inhabitants. Despite the fact that only 19 of the 87 resorts in the country had to close due to physical damage, it is thought that the total losses of this sector accounted for 60 or 70 % of all negative effects for the country (RoM 2005:20). Because of the country's dependency on income from tourism, much effort was made to reopen the resorts after a relatively short time.

In literature on tourism marketing, 'place perception' plays an important role in the development of sales strategies as well as in coping with negative effects of hazardous events (SANTANA 2004). In the case of the impact of hazards, several other factors besides the actual physical damage resulting from a stigmatisation of this place have to be taken into account when attempting to explain the recovery period and the strategies which can be implemented to tackle the decline in visitors. ICHINOSAWA (2006) developed a framework on this basis to analyse the recovery period of the tourism sector in the Phuket Province (cp. Figure 28).

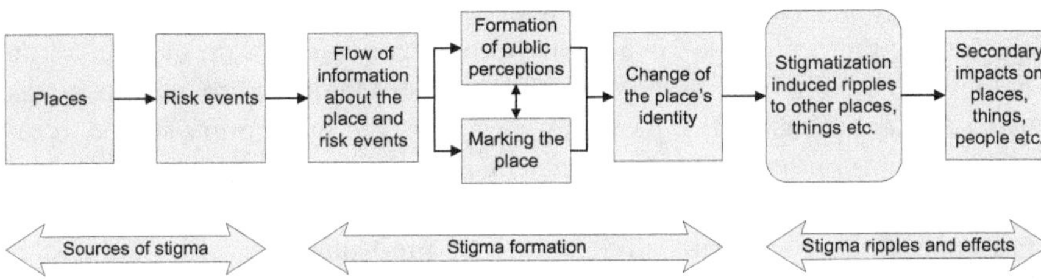

Figure 28: Process of place perception after hazard impact (Author's design, based on ICHINOSAWA 2006:116)

The process of changing the public perception of the tourism destination is strongly affected by the media coverage of the event, including pictures of the devastation and the personal incidents. Due to the large number of foreigners involved in the catastrophe, the media coverage as well as the information recorded by the tourists themselves and broadcasted by the media received great attention worldwide. Additionally, ICHINOSAWA found several media reports doubting the local authority's' capabilities of dealing with the situation. This influence on the perception of place in the origin countries of the main tourist markets for Phuket is reinforced by the fact that unfamiliar or distant places may easily lead to distorted or stereotypical perceptions, which is particularly the case for a long-haul destination such as Thailand or the Maldives (KASPERSON et al. 2004:12). Due to this unfamiliarity, the public is not able to determine where pictures were taken and to distinguish between the nearly completely destroyed area of Khao Lak and the mostly untouched island of Phuket. As a result, Phuket Island was marked as a tsunami catastrophe zone in the perception of the potential tourists due to the pictures from Khao Lak, with the consequences of a tourism decline and an economic stagnation.

In order to prevent such a public stigmatisation, the Maldives Tourism Promotion Board attempted to develop a strategy to transfer a positive image of the island risk management capacities to the public media. Firstly, the strategy aimed to stop short-term cancellations and to send out a picture of a rapid response to the devastation by starting immediate reconstruction. To promote this image, direct contact to travel agencies and charter airlines in the target markets was established to convince the sales representatives of the effective measures taken in the Maldives. Additionally, financial support was granted in order to prevent the resorts from giving discounts which could lead to the consumers' perception of a problematic situation with a reduced number of tourists in the destination country, but also have a long-term negative effect on the reputation of a high-price long-haul destination, which the Maldives are positioned as (MIDDLETON et al. 2009:341). As a long term strategy, the country started to reduce the reliance on travel intermediaries by participating in tourism exhibitions and accessing new markets such as China or India. As a consequence, the country's occupancy levels had already reached the pre-tsunami level by the end of 2005 (CARLSEN and HUGHES 2008; Figure 29).

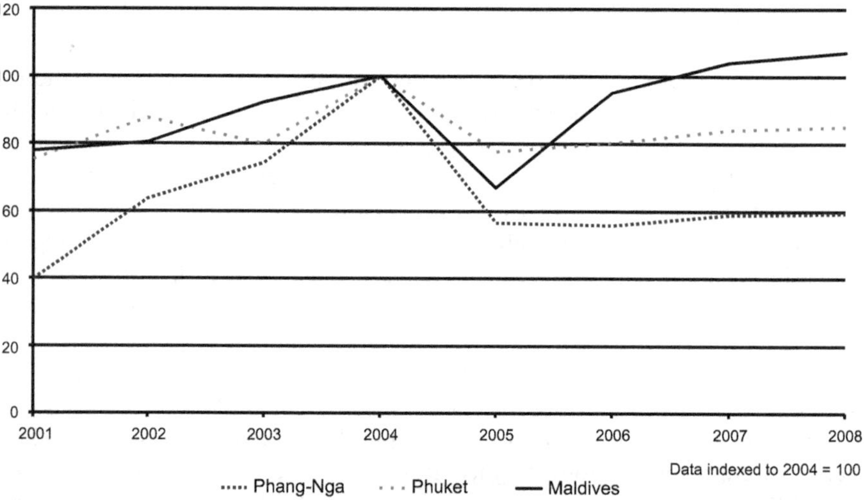

Figure 29: Development of value addition in the 'Hotel and Restaurant Sector' compared to 2004 (Author's design, based on data from NESDB 2009; MPNP 2009)

In contrast, Thailand's tourism regions experienced a much longer recovery period which, in some cases, still has not reached the extraordinary results of 2004. The promotion of tourism activities in Thailand is concentrated in the government-owned TAT, which was founded in the 1960s (TAT 2009a). In the past decade, the tourism sector in Thailand has had to face several events dampening the positive development in occupancy rates. These started with the terrorist attacks on 11 September 2001 and continued with SARS, the political crisis and the following coup d'état in 2006, the political instability which followed, and the worldwide economic crisis starting in 2007. As a reaction to the tsunami, the TAT launched the campaign and recovery package "Andaman Recovery" with the slogan "Andaman Smiles Again", including the "Lodtangkoh Sanuktangmuang" (fun package), consisting of a return trip, accommodation and transfers for a relatively low price. Additionally, to promote domestic tourism, duty-free shopping for overnight guests from Thailand in the provinces of Phang-Nga, Phuket and Krabi was initiated. On the company level, strategies were also developed for restarting business activities. More than 40 % of the tourism-related businesses stated that they had changed their business strategy or initiated promotion packages for their customers. However, only 20 % said that they had attempted to improve their business activities by extending their service or offering new products to their customers. Nevertheless, more than 60 % of all companies which experienced the indirect effects of the natural hazard stated that they had not changed anything at all in their business activities.

Despite this, domestic tourism was not able to cover the reduction in international arrivals due to the religious and traditional constraints of the Buddhists. In this religion, people believe that when a person dies, the body has to be buried or cremated by a relative. If this is not done, or if the body is not found, it is believed that the deceased will appear again and again in this specific location until religious rites are performed (RITTICHAINUWAT 2011). Buddhists from China, Taiwan, Hong Kong, Singapore and Thailand believe that these 'ghosts' can cause misfortune as well as physical or mental illness (BURNARD et al. 2006:743). Since a large number of corpses of tsunami victims were washed out to sea, and therefore have never been found and buried or cremated, the Andaman coastline is still considered a 'bad luck location' that might cause 'ghost illness' (HUANG et al. 2008). Therefore, most Asian tourists changed their destinations to alternative places in the Gulf of Thailand, for example Hua Hin or the Eastern Seaboard.

The first returning tourists to the area were repeat tourists and those who had close contacts to people living in the tsunami-hit areas. These people had a better understanding and a clearer picture about the extent of the damage and the progress of the recovery activities than was actually portrayed by the media (RITTICHAINUWAT 2006). Around one year after the event, RITTICHAINUWAT (2006:396) found that more than half of the visitors in the Phuket Province were repeat travellers, which was considerably higher than in the following years. These people were interviewed about their decisive factor for travelling to the

Andaman Coast. Surprisingly, the first eight reasons given by the interviewees stressed characteristics of the country which had also been highlighted in the past TAT campaigns, for example cultural heritage, culinary delights, hospitality, nature and climate (McDowall and Wang 2009:356). The high score relating to the answer category 'good value for money' is a traditional strength of the Thai tourism sector (Dwyer 2000; McDowall and Wang 2009:360). Tsunami-related topics, such as 'Image of Thais helping foreigners during the tsunami', 'Installation of tsunami warning system' or the 'Curiosity to see the improvement/debris of the disaster' were low on the list and also showed the highest standard deviations, which points to an inconsistent answering pattern. Particularly outstanding is the low interest and motivation to visit the area based on the 'Low-cost tour package', which had been specially designed to motivate tourists to return to the area.

Underlining this fact, Rittichainuwat (2006:400) cited a statement of the chairman of the Phuket Chamber of Commerce, who said that "several hotels in Phuket have been overbooked since November 2005 because of aggressive marketing efforts, such as road shows and the support from tour operators and loyal guests, but not from the low-cost tour package." Consequently, the official TAT statistics report the drastic decrease of revenue per guest for the years 2005 and 2006 in the Phuket Province, and also a slight drop in revenue per person for the Phang-Nga Province (cp. Table 22). In the Phuket Province, the number of visitors reached the level of 2004 three years after the impact, meaning that the total revenue level of the pre-tsunami time could be achieved again. In the Phang-Nga Province, however, the level of 2004 in terms of the number of visitors and the total revenue has still not been reached. One reason for this is the drastic over-capacity in the Phang-Nga Province, where the number of rooms grew between 2003 and 2007 by around 61 % despite the tsunami devastation. During this period, the Phuket Province experienced an increase in room capacities of around 19 %, which was absorbed by the rising numbers of guests. As a consequence, the occupancy rate in the Phang-Nga Province dropped from 51 % to around one-third, while it could gain 7.4 percentage points to reach 66 % in the Phuket Province. Besides the drastic decrease in the number of visitors in Phang-Nga, the low expenditure per day and per person also seems to have been a serious problem for the province's tourism business. While in the strongest year of 2004, the average expenses per person and day reached 83 % of the average for the whole kingdom, the percentage in the following years remained on a slightly lower level of 80 %. In the Phuket Province, however, the average expenditure per person and day reached more than four and a half times the level of the nation's average (TAT 2007b). The reason for these different average expenditures is the strongly tourism-orientated surroundings on Phuket Island offering outdoor parks, shopping malls and other leisure and entertainment activities, which are mostly unavailable in the Phang-Nga Province (FG-Hot-PB). This fact can also be underlined by the occupancy rate, which stays at around 70 % all year in the Phuket Province, while in Phang-Nga, it ranges from 60 % in the high season, which is relatively cold and dry, to 15 % in the hot and rainy season.

Table 22: Development of tourism in Phang-Nga and Phuket Province (Author's own design, based on TAT 2007a; TAT 2007b; TAT 2009b; TAT 2009d)

Number of visitors	2002	2003	2004	2005	2006	2007	2008
Phang-Nga	80.4	80.7	100.0	28.4	35.3	40.1	46.4
Phuket	83.3	84.5	100.0	52.4	93.9	104.4	110.8
Revenue							
Phang-Nga	70.0	65.5	100.0	26.4	34.2	39.9	43.2
Phuket	84.7	85.5	100.0	32.9	90.6	110.0	118.7
Revenue per Person							
Phang-Nga	87.0	81.2	100.0	93.1	96.8	99.6	93.2
Phuket	101.8	101.2	100.0	62.8	96.5	105.3	107.1
Data indexed to 2004 = 100							

Knowing these facts and facing the over-capacities and low revenue, the Phang-Nga Tourism Association and the local business association of Khao Lak turned their attention to new customer groups and to promoting new ideas for the development of the area. Prior to the tsunami, the growth strategy had attempted to follow the example of Phuket, and more and more international hotel chains started to build resorts in Khao Lak, some of which were finished shortly after 2004 (TRAIT Company Survey; CALGARO and LLOYD 2008:294). In the following years, Khao Lak turned to an extremely price-sensitive market but still relied on high-standard four or five star hotel resorts. The customer group of this area, with around one-third being returning guests and most of them staying for more than 10 days, has a disposable income that is significantly higher than for other parts of Phang-Nga or even Phuket (INTERVIEW KHAO LAK HOTEL ASSOCIATION). The aim was to keep this customer group in Khao Lak by focusing on the slogan "your personal paradise", which implies a vision of not locating more resorts in the area, but of spreading out from Khao Lak to the surrounding islands or regions to maintain a low-density tourist area with a strong focus on peaceful tourism. Additionally, the level of night life entertainment in the form of bars, shows and other entertainment was to be kept to a minimum level. To increase the expenditure volume of the tourists in the area, a strong emphasis was placed on nature-oriented tours in the nearby Khao Sok national park. Therefore, tour agencies and resorts were likely to cooperate to a stronger degree than now. This cooperation between the local business owners had been described as very strong by the President of the Hotel Association Khao Lak, since most of the owners are local people who meet on a daily basis for leisure activities and to talk and discuss the development of the area. This statement has also been confirmed by other companies. This unique collaboration between the key stake holders in this area might be a strong advantage for further development in the area. For Patong Beach, no collaboration between the stake holders on the company level could be confirmed, and instead, a strong rivalry was mentioned between local business owners and investors from other parts of the kingdom or foreign countries, which were described as not being interested in local development (INTERVIEW PHUKET MERLIN HOTEL).

Therefore, the situation in Phang-Nga five years after the tsunami can be summarised by stating that the number of domestic tourists is dampened by the remaining effects of the tragedy, which might be one major factor reducing the proportion of visitors from Buddhist countries from 60 % to less than 50 %. On the other hand, it was possible to increase the number of visitors from high-income countries such as Central Europe and Scandinavia, but their expenses per person and day still remain significantly below the nation's average and are far beneath the expenditure level of Phuket. The reason for this low income level, besides the underdeveloped tourist infrastructure, is the over-capacity and the resulting low occupancy rates for hotels and resort rooms. The Phuket Province, on the other hand, was able to achieve its revenue level within two years after the disaster and still maintains the strong position in terms of the expenditure level of the tourists. As well as the high percentage of visitors coming from high income countries such as Australia, the United Kingdom, Sweden, Korea or Germany, the percentage of visitors in the last year coming from Russia and Malaysia has increased drastically to more than three times the level of 2003 (cp. Table 23).

Table 23: Home country of tourists (Own calculations based on TAT 2004a, TAT 2004b, TAT 2007a, TAT 2008b, World Bank 2011)

	Phuket		**Phang-Nga**	
	2003	**2007**	**2003**	**2007**
High income	59.62%	58.99%	30.76%	49.10%
Upper middle income	1.69%	6.01%	0.62%	0.31%
Lower middle income	30.22%	32.45%	60.50%	48.31%
Other	8.46%	2.55%	8.12%	2.29%

Visitors from these two emerging markets are almost completely absent in the Phang-Nga Province, and the local tourist association informally agreed not to open up to markets such as Russia or Arabia due to different needs in services compared to the Central European or Scandinavian customers (ATILGAN et al. 2003; INTERVIEW KHAO LAK HOTEL ASSOCIATION).

7 Causal factors for vulnerability on the micro-level

As derived from the theoretical discussion in Chapter 3 the effects of a natural hazard on an economic system are determined by the factors of exposure and sensitivity. As shown in Figure 11, exposure and sensitivity in economic systems are formed by factors deriving from household and company characteristics and the economic setting on the regional level. The third influencing factor of resilience mostly influences the ability of the system to cope with and to recover from the impact (THYWISSEN et al. 2006). The factors likely to support the coping and recovery capabilities will be discussed in this chapter. These factors will be further investigated in the study areas of Khao Lak, Ban Nam Khem and Thai Muang in the Phang-Nga Province, as well as in Patong Beach in the Phuket Province (cp. Map 7).

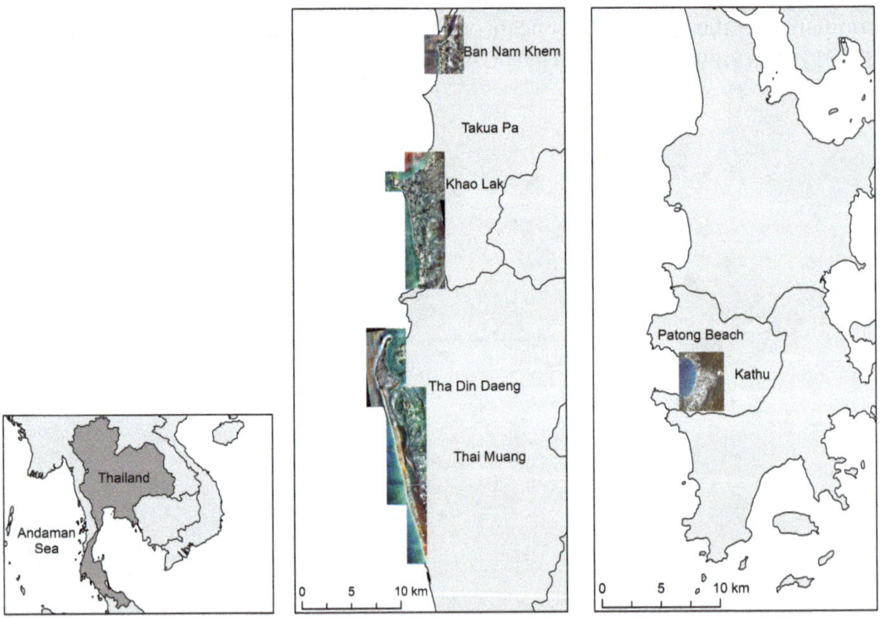

Map 7: Study areas in Phang-Nga Province (left) and Phuket Province (right) based on IKONOS multispectral data from January 2003 (Author's own design)

7.1 Representativeness of study areas for southern Thailand

These four communities were chosen for representative usage along the Thai Andaman Coastline. This requirement must be fulfilled for the economic structure as well as for the building structure. The representativeness of the building structure was tested using the Chi-square test with the hypothesis H_0 that the two samples - the southern provinces of Thailand and the study communities altogether - are from identical basic populations. The Chi-square values were under the threshold for rejecting the H_0 hypothesis with a 5 % significance level (BACKHAUS 2008:241). This shows that the results from field mapping and the household survey are from the same basic population. In a second step, the results were tested in relation to the housing characteristics stated by a socio-economic survey for southern Thailand. Differences were found in Patong Beach, with its highly urbanised building structure, showing a lower share of detached houses than in rural areas but with more row houses and households living in apartment blocks. In contrast, the community of Thai Muang is very rural and almost exclusively shows a building structure dominated by detached houses, which differs from the average in southern Thailand. The four surveyed communities as a whole show a significant similarity to the data provided by the NSO for southern Thailand. This study can thus be called representative for the study area and southern Thailand in terms of the building structure (cp. Table 24).

Table 24: Representativeness of building characteristics (Own calculations based on TRAIT household survey and data for southern Thailand by NSO 2007e)

	Housing characteristics			X^2-square
	Detached	**Row House**	**Flat**	**Southern Thailand**
Ban Nam Khem	88.3 %	4.9 %	6.8 %	13.08
Thai Muang	98.1 %	0.0 %	1.9 %	21.11*
Khao Lak	61.5 %	33.8 %	4.6 %	8.85
Patong	56.0 %	32.1 %	11.9 %	16.74*
Study communities	73.2 %	17.1 %	9.7 %	7.38
Southern Thailand*	79.5 %	18.9 %	1.0 %	--

* 2-side significant different at 1 % - n = 323

In addition to the representativeness according to housing characteristics, socio-demographic characteristics are also an important influence on vulnerability. The household income in particular is a strong indicator for a variety of socio-economic household characteristics, and the population assessed in the household survey should be tested for its comparability with the data representative for southern Thailand. The households that were surveyed by the TRAIT project show a distribution similar to the one in southern Thailand. The chi-square test generated very small values, indicating a close relation of the distribution. Only the households surveyed in Patong Beach differ from the distribution in southern Thailand as indicated by the Chi-square Value of 27.4. This is due to the strong influence of the flow of external spending capacity into the city by the strong tourism sector, which leads to the fact that the percentage of households in the higher income classes is above average (cp. Table 25).

Table 25: Monthly household income in Euro for 2008 (Own calculations based on TRAIT household survey and data for southern Thailand by NSO 2007e)

	<65	**65-110**	**110-215**	**215-325**	**325-650**	**650–1,080**	**> 1,080**	**X^2 compared to Southern Region**
Ban Nam Khem	5.9%	5.9%	28.4%	17.6%	24.5%	10.8%	6.9%	1.4
Thai Muang	5.7%	7.5%	28.3%	17.0%	32.1%	7.5%	1.9%	4.5
Khao Lak	3.0%	4.5%	24.2%	12.1%	31.8%	13.6%	10.6%	4.2
Patong Beach	5.1%	1.0%	9.1%	8.1%	35.4%	20.2%	21.2%	27.4*
Study area	5.0%	4.4%	21.6%	13.4%	30.6%	13.8%	11.3%	3.9
Southern region**	3.9%	6.6%	23.6%	20.1%	27.7%	12.1%	5.9%	

* 2-side significantly different at 1 % n = 323; ** Data for 2007

The employment structure of the surveyed household members in the different business sectors shows a similar distribution to the one in southern Thailand assessed by the National Statistical Office. The Chi-Square Test advises accepting the H_0 hypothesis that the two distributions result from the same basic population with a significance of 1 %. In the southern provinces, half of the workforce is still employed in the agriculture and fishing sector, while in the study areas, this share is limited to a quarter of the total workforce. The manufacturing and construction sector shows a similar proportion of the workforce, with 14 % in the southern provinces and 11 % in the study area. Because of its high importance of generating external income flows, the service sector has been split into three different sectors in this investigation, namely the wholesale and retail sector, the hotel and restaurant sector, and other services. In each of the sectors 'wholesale and retail' and 'hotel and restaurant', 20 % of the workforce of the study area is employed. 'Other services' accounts for 25 % of the employment of the workforce in the study region. This distribution of employment shows a specialisation of the region towards the tourism-oriented service sector, especially in Khao Lak and Patong Beach. The chi-square test indicates that it is necessary to reject the H_0 hypothesis for these two communities, while Ban Nam Khem and Thai Muang show a highly significant similarity in the distribution of the workforce. Overall, the four studied communities form a region that gives a good representation of the economic situation in southern Thailand.

The sampling of companies was conducted in relation to the total population assessed in the field mapping according to the 16 sector TSIC classification. For the three study communities in Phang-Nga (Khao Lak, Ban Nam Khem and Thai Muang), the official database from the Department of Business Development and additional data from the regional office of the NSO refer to 17 businesses in total in the whole region. For Patong Beach in the Phuket Province, these databases offer a similar number of entries (DBP 2011). The number of informal, non-registered economic activities exceeds the number of official business activities in this region by a huge amount. Statistics for the economic sector only cover certain sectors such as the industrial sector or the service sector, but are not linkable in terms of time and regional scale. Despite this, the companies mentioned in the official databases were interviewed with increased preference. Comparing the interview data to the statistical region of southern Thailand in terms of the number of companies per business sector is not possible because official and complete data on this information is not available. When using the Chi-Square Test to compare the distribution of businesses within the sector, there is a strong relation between the company survey and the results from the field mapping (cp. Table 26).

Table 26: Share of persons working in interviewed companies including owner and family members (Own calculations based on TRAIT company survey and data for southern Thailand by NSO 2007d)

	Agriculture and fishing	**Manufacturing and construction**	**Wholesale and retail**	**Hotel and restaurant**	**Services**	**Number of companies interviewed**
Ban Nam Khem	9.2%	35.1%	40.5%	10.7%	4.6%	39
Thai Muang	47.4%	2.6%	31.6%	15.8%	2.6%	19
Khao Lak	1.5%	26.2%	14.2%	37.5%	20.7%	66
Patong Beach*	0.0%	2.2%	17.6%	52.8%	27.4%	172
Study Area*	2.9%	11.6%	19.8%	43.2%	22.4%	296
Phang-Nga	22.8%	55.9%	8.2%	4.1%	8.9%	
Phuket	4.4%	8.8%	23.5%	34.9%	28.4%	
Phang-Nga and Phuket	16.8%	40.6%	13.2%	14.2%	15.2%	

* Chi-Square = 15,36 (Study Sites in Phang-Nga and Phuket Province; Patong Beach and Phuket = 12,83 - n = 296

In addition to the validation of the survey with the field mapping results, secondary statistics about the employment figures in the different business sectors provided by the different regional offices of the NSO were also used. Information about the distribution of companies in terms of absolute numbers of companies is not available from the National Statistical Office. The company structure found in the four study communities differs greatly from the aggregated employment situation in the provinces of Phang-Nga and Phuket. The Chi-Square Test rejects the H_0 hypothesis with a strong significance. Furthermore, it indicates that the study area along the coastline of the Andaman Sea is comparable with the employment situation in the province of Phuket, with a strong focal point on the tourism-related business sectors such as 'hotels and restaurants' and 'retail trade'[2]. Only in the communities of Ban Nam Khem and Thai Muang do agriculture and fishing still play a significant role, although the service sector is already highly important. In all three communities surveyed in the Phang-Nga

2 Due to the restrictions in the chi-square test not tolerating cell values lower than five, the hotel and restaurant sector have been aggregated for the evaluation of the representativeness (BACKHAUS 2008:241pp).

Province, manufacturing and construction is of minor importance compared to the numbers for the province of Phang-Nga. Because the three communities in Phang-Nga are located close together within a range of 15 km, they are strongly linked and can be aggregated and investigated as a common economic region. The distribution of the workforce among the business sectors for each of the communities is comparable neither to the Phuket Province nor the Phang-Nga Province. Taken as a whole, the distribution here is significantly similar to the Phuket Province. Also, the economic situation in Patong Beach shows a similar distribution to the total economic situation in the Phuket Province.

Therefore, it can be concluded that the economic situation in the four communities is highly influenced by the tourism sector. While the economy in Patong Beach is already strongly focused on this income source, the communities in the Phang-Nga Province tend to depend on this sector, which might become more important in the coming years. While the household survey in the communities of Phang-Nga is representative for all of the southern provinces of Thailand in terms of employment situation and income situation, the businesses located in these communities differ from southern Thailand, and are more oriented towards the Phuket Province. This indicates that not all the workforce is employed in local businesses, but also commutes to other nearby cities such as Takua Pa.

7.2 Causal factors determining tsunami effects in the communities

The four communities represent the special economic setting of southern Thailand. Khao Lak itself constitutes a booming tourist hotspot at the centre of a region that is still dominated by agriculture and fishing. Since the 1990s, Ban Nam Khem, a former tin mining community has been economically dominated by fishing and aquaculture. Situated south of Khao Lak on a peninsula, the community of Thai Muang depends mostly on small-scale agriculture. Patong Beach in the Phuket Province represents a developed hub for international tourism in southern Thailand. In all of these four communities, household and company surveys as well as qualitative investigations for assessing the mitigation strategies after the disaster were conducted. Since panel data for these areas were not available, the households and companies were interviewed about their current situation in the year 2009 and asked if changes had occurred since the time directly before the tsunami. Households or companies that had changed something in their livelihood or business situation were not included in the investigations for the situation before the tsunami. In the case of households, 16 % had changed something in their working situation and were therefore not included in the further investigations. In terms of their housing situation, 12 % had changed their location or the characteristics of their building, so they were also excluded from the subsequent investigation of the pre-tsunami situation. In the case of companies, one quarter had changed their business strategy fundamentally, so were not included in the analysis of the pre-tsunami situation.

7.2.1 Creating the database for household and company exposure

Assessing the economic exposure of the study areas means investigating the assets that are potentially at risk. Therefore, the totality of physical assets available in companies and households needs to be estimated despite the lack of official data on the chosen micro-level. Since the economic consequences of a natural hazard are not only limited to physical damage, but also spread out along the connections within the economic system between households and companies or between companies, the flows of labour and goods are included in the assessment as well (cp. Chapter 3.2.1.2).

For the residential buildings, a linear regression model was developed. This regression model operates on dwelling units and combines physical characteristics of the building with image pixel characteristics as well as information from the household survey. Indicators were derived using manually digitised building shapes and the automatically extracted building shapes for the years 2003 and 2008 in the Phang-Nga Province. From this and the underlying remote sensing pictures, indicators about building size, roof characteristics and location could be derived. For the more rural areas of Ban Nam Khem, Khao Lak and Thai Muang, the use of the colour of the roof as a predictor variable showed a significant improvement in the estimations. Therefore, a binary variable was created indicating a grey roof. For the urban area of Patong Beach, this indicator showed no significant influence due to the urban building structure with its flat-roofed apartment complexes. Location information in terms of distance from coastline or height, which would be able to describe neighbourhood relations as well as specific quarters, showed no significant influence, which indicates that no segregated residential areas can be found in the study areas. For the communities in the Phang-Nga Province as a whole, a regression model was set up showing significant influences of roof colour and building size (cp. Table 27).

Table 27: Linear regression model for the prediction of people distribution (Own calculations based on TRAIT household survey)

	Number of household members permanently living in household	
	Patong Beach	**Phang-Nga communities**
Constant	3.18	2.333
Size of house in m^2	0.01 (0.403)	0.013 (0.331)
Grey roof	ns (ns)	0.997 (0.349)
corr. R^2	0.153	0.202
significant at 5 % error level; beta values in brackets; n = 323		

The application of the regression model has an advantage over the usage of the procedure of distributing the number of inhabitants according to the indicator 'persons per square meter' in that building characteristics are included. The indicator 'grey roof' in particular has a correctional influence on the relation between the 'number of persons living permanently in the household' and the 'size of the building'. The correlation between these two indicators shows a significant dependency when it is controlled by the indicator 'grey roof'. The indicator 'grey roof' shows a significant correlation of -0.22 with the 'household income' and of -0.26 with the 'size of the house'. Grey roofs indicate roofing made of corrugated iron sheeting or roofing felt, which is cheaper than coloured roof tiles. This dependency can be found in the regression model because for households which have a roof that is not grey, the number of persons is lowered by nearly one. This is supported by the significant correlation of the 'average income per household member' with the 'number of persons permanently living in the household' of – 0.21.

The outcome of the regression model can be tested on the municipality of Phuket. Since this community is the only study area matching an official administration unit, official data are available here which can be used for comparison. The application of the regression model for Patong shows an acceptable estimation of the total population for this area. Official numbers for the municipality of Patong from the NSO for 2003 report nearly 15,000 inhabitants in more than 9,000 households (NSO 2004). The estimation using the regression model calculates a number of persons living in Patong of close to 20,000 in nearly 3,000 houses, which overestimates the official numbers by 30 %. Due to the fact that the regression model is based on the household survey with its sampling using the field mapping, it can be assumed that the survey represents the actual or 'de facto' population, which means the population that is physically represented in the area. The official number of inhabitants is based on the official population register, which represents the 'de jure' population. Since Phuket shows an inward migration ratio two times higher than the outward migration rate, it can be assumed that a large number of 'de facto' residents are not registered in the area. Compared to the differences between the official number of inhabitants for the Bangkok Metropolitan Area of around 6 million and the estimated numbers of 9 million (KRAAS 1996), the regression model for Patong Beach gives an acceptable result for the distribution and the number of the total population of the municipality. For a better interpretation of the distribution, the results were aggregated to a 100 m² grid.

By applying the regression model developed for the Phang-Nga Province to the result of the automatic building extraction procedure, a total number of 4,700 persons resulted for Ban Nam Khem, which corresponds to the findings of TIFFERT (2010). The average household size is 3.77. The population of Ban Nam Khem lives almost entirely inside the area which was inundated by the tsunami waves in 2004. Official numbers for Ban Nam Khem are not available, since the numbers are only available on the level of municipalities (Amphoes). Prior to the tsunami, 1,566 households with 4,100 villagers were counted (PHAPHAVASIT et al. 2009:22).

Building upon the number of household members distributed to each of the buildings in the study region, the household income can be calculated for each household. The independent indicators all show a significant influence for the dependent variable, except the indicator 'grey roof', which is not significant in Patong Beach, and therefore excluded from the regression model. The R^2 values in both of the areas only reach a 'weak' explanation of 0.15 and 0.12 respectively, which must be accepted since all indicators that can be derived from remote sensing are included in the model. The quality of the model could only be extended by using other supporting variables, namely secondary data on the household level, i.e. geo-marketing data, or official data on the household level (MEYER 2005; REESE 2003). The application of this kind of data is not the intention of this study, and therefore the regression model is accepted. The regression models of Patong Beach and the study communities in Phang-Nga show obvious differences. In Patong Beach, the size of the house shows a significant influence, with the number of household members also showing a strong influence on the total household income. In contrast to this, the constant value in the Phang-Nga communities is much higher and the additional influence from the number of household members is clearly lower. Furthermore, the 'grey roof' indicator shows a significant influence, lowering the household income. The household income in the Phang-Nga study area shows a mean value of nearly 20,500 THB (450 €), while in Patong Beach, the average income is nearly double that, at almost 39,000 THB (850 €). This disparity expresses itself in the different constant values and the multiplier for each household member. Another point is the indicator 'size of the house', which is not significant for Phang-Nga communities, but is significant in Patong Beach (cp. Table 28).

Table 28: Regression model for household income (Own calculations based on TRAIT household survey)

	Household income	
	Patong Beach	**Phang-Nga communities**
Constant	16,543.7	25,528.1
Size of house in m²	79.60 (0.37)	ns (ns)
Grey roof	ns (ns)	-17,858.9 (-0.32)
Number of household members	3,159.4 (0.15)	2,302.8 (0.18)
corr. R^2	0.15	0.12
significant at 5 % error level – beta values in brackets – n = 323		

In Phang-Nga, no significant correlation could be found between the number of household members and the size of the house which stands in contrast to Patong Beach. This difference can be substituted through the variable 'grey roof' which indicates a lower income level (cp. Table 28). Using a partial correlation controlled by this indicator gives a significant correlation between the number of household members and the size of the house, which leads to the assumption that larger households with a lower income in particular live in houses that have roofing made of corrugated iron sheet or roofing felt.

In the next step, the household assets for each of the households are calculated. The indicator and the constant value show the same differences between Patong Beach and the Phang-Nga communities mentioned above (cp. Table 29). This results from the fact that the amount of assets, which includes property value, long-term consumable assets and vehicles, is nearly twice as high in Patong Beach, with 12,326 €, as in the Phang-Nga communities, where the average amount of household assets is 6,526 €.

Table 29: Regression model for household assets (Own calculations based on TRAIT household survey)

	Assets including vehicles and property	
	Patong Beach	**Phang-Nga Communities**
Constant	276,657.4	189,392.6
Size of house	1,484.9 (0.24)	1,388.3 (0.51)
Grey roof	ns (ns)	-62,650.7 (-0.10)
Total household income	4.61 (0.24)	2.58 (0.30)
corr. R^2	0.25	0.38
significant at 5 % error level – beta values in brackets – n = 323		

In the last step, the distribution of expenditures is calculated. No significant influence of building characteristics can be identified here either. Nevertheless, the household expenditures are connected to the household income, and the relation can be described by an exponential function. The average expenditure per household in Patong Beach is nearly 19,000 THB (415 €), and in Phang-Nga 13,000 THB (290 €), which matches the findings of the NSO for the Phuket Province (NSO 2007e). For the Phang-Nga Province, the average expenditure extends the provincial mean value of 9,962 THB, which refers again to the better economic situation in the study regions caused by the tourism-generated income (NSO 2007e). The differences in income and expenditure between the two provinces can also be found in the regression models with the differing constant values (cp. Table 30).

Table 30: Regression model for household expenditure (Own calculations based on TRAIT household survey)

	Household expenditure	
	Patong Beach	**Phang-Nga communities**
Constant	11649.3	10071.9
Household income	$7.53\ 10^{-6}$ (0.50)	$6.56\ 10^{-6}$ (0.49)
corr. R^2	0.247	0.242
significant at 5 % error level – beta values in brackets – exponential function - n = 323		

On the company side, directly and indirectly exposed elements - following the argumentation of SMITH and WARD (1998:35) - were included in the assessment. On the direct exposure side, property value and productive assets were evaluated. In the case of indirectly exposed elements, financial flows and trade flows were integrated. Due to the limited sample size and the differences among the six differentiated business sectors and the two provinces, it was not possible to create a similar set of regression models in order to predict directly and indirectly exposed elements on the company side. Therefore, the physical assets and the sales levels were allocated to the complete group of companies by creating percentile classes based on the survey data. These percentile classes could be linked to significantly differing building sizes divided into the provinces and then designated to each of the identified companies. In order to identify the interregional linkages of the companies, the cost structure of the six business sectors was derived from the official input-output table (cp. Table 31) and linked with the companies' statements about the origin of their supply and their targeted markets.

Table 31: Cost structure of companies 2005 (Own calculations, based on data by NESDB 2006)

	Agriculture and fishing	**Manufacturing**	**Wholesale and retail**	**Restaurants**	**Services**	**Hotels**
Agriculture and fishing	8.6%	7.6%	0.0%	16.0%	0.6%	1.3%
Manufacturing	26.9%	58.5%	7.1%	41.2%	27.5%	43.0%
Wholesale and retail	0%	0%	0%	0%	0%	0%
Restaurants	0.0%	0.1%	2.4%	0.1%	0.4%	0.3%
Services	0.1%	0.1%	0.3%	0.0%	0.3%	0.1%
Hotels	2.3%	3.2%	8.7%	1.5%	11.5%	11.5%
Wages	15.9%	8.6%	13.3%	9.4%	30.5%	16.1%
Surplus	41.6%	13.8%	55.2%	18.8%	16.9%	16.9%
Depreciation	4.6%	5.0%	7.5%	6.2%	10.0%	4.3%
Taxes	0.1%	3.2%	5.5%	6.9%	2.3%	6.6%

Paying sector in rows, receiving sector in columns

7.2.2 Evaluation of exposure on the micro-level

The community area of Ban Nam Khem is highly exposed to natural hazards due to its flat terrain and its proximity to the coastline. During the tsunami of 2004, the whole area of the community was inundated by the waves. All buildings in the area were affected by the wave impact, and only a minority of roughly 50 buildings were not destroyed by the impact. By applying the regression model developed for the Phang-Nga Province to the result of the automatic building extraction procedure, a total number of 4,700 persons resulted for Ban Nam Khem. The population of Ban Nam Khem lives almost entirely inside the area which was inundated by the tsunami waves in 2004. A common method of estimating the death toll is to apply an empirically based equation to the household which sets the death toll in relation to the number of persons and their height above sea level (HARBITZ et al. 2011; REESE et al. 2007). Applying a formula developed by KATHIRESAN and RAJENDRAN (2005) which calculates the number of personal casualties as a dependency of height above sea level, an estimated death toll for the community of Ban Nam Khem of 1,505, with a mean value of 2.3 per household, was calculated. During the tsunami of 2004, two members of the affected households of the community were killed on average by the waves (PHAPHAVASIT et al. 2009). The official numbers for Ban Nam Khem from the DDPM give a total death toll of 1,994 from the wave impact of the tsunami (DDPM 2005b).The economic assets exposed to natural hazards per household are relatively high compared to the three other study communities (cp. Table 32; Table 33).

Table 32: Average exposure key facts per household of the study communities 2004 (Own calculations based on TRAIT household survey)

	Total assets €	Total income in €	Total expenditures in €
Ban Nam Khem	$12{,}600_{a}$	407_{a}	322_{a}
Thai Muang	$5{,}020_{a}$	358_{a}	218_{a}
Khao Lak	$7{,}952_{a}$	$558_{a,b}$	544_{a}
Patong Beach	$14{,}531_{b}$	827_{b}	404_{a}

Subscripts in the same row not sharing the same letters are significantly different at $p<0.05$ in the two-sided t-test – n = 171

For the year prior to the tsunami, two-thirds of the interviewed households reported owning the property on which they live, while in Khao Lak and Patong Beach, less than half of the households owned the property where they lived. As a result of the relatively high concentration of assets per household, the total amount of exposed assets in Ban Nam Khem nearly reaches the share of the exposed households assets located inside the inundation area of Khao Lak (cp. Table 37; Map 8). Around one-tenth of the financial income of the households in the community is generated from transfer payments from government sources, such as elderly assistance or scholarships, and transfer payments of household members living apart from the households and sending remittances to their family. These income sources can be seen as mostly not endangered by regionally limited natural hazards such as the tsunami. The majority of income for the population in Ban Nam Khem is generated through wages and salaries or income from self-employment. The workplaces of the population in Ban Nam Khem are mostly concentrated within the community; around 70 % of the household income is generated from work in Ban Nam Khem. 10 % of the workforce is employed in the area of Khao Lak, and only a marginal share of the population works in other nearby communities. Since the majority of people work for companies located inside the inundation area of the 2004 tsunami, the income sources can be deemed as highly exposed to natural hazards.

Table 33: Exposure of households in Ban Nam Khem 2004 (Extrapolation based on TRAIT household survey data and automatic building extraction from MFC data)

	Total	**Thereof located in inundation zone**
Number of person	4,656	98.6%
Amount of assets in 1,000 €	8,600.8	98.6%
Amount of income per month in 1,000 €	357.5	98.5%
Amount of expenditures per month in 1,000 €	166.7	99.0%

The majority of the expenditures of the population are also concentrated within the community. Goods for the short and mid-term needs in particular are purchased in stores in the community or in the nearby market in Bang Muang. Long-term durable goods, such as household appliances, furniture, electronics and vehicles, are mostly purchased in the district capital Takua Pa, located around 10 kilometres inland. These long-term consumable assets were purchased mostly after the tsunami in 2004, but looking at the households not affected by the tsunami shows that these items are usually purchased every couple of years.

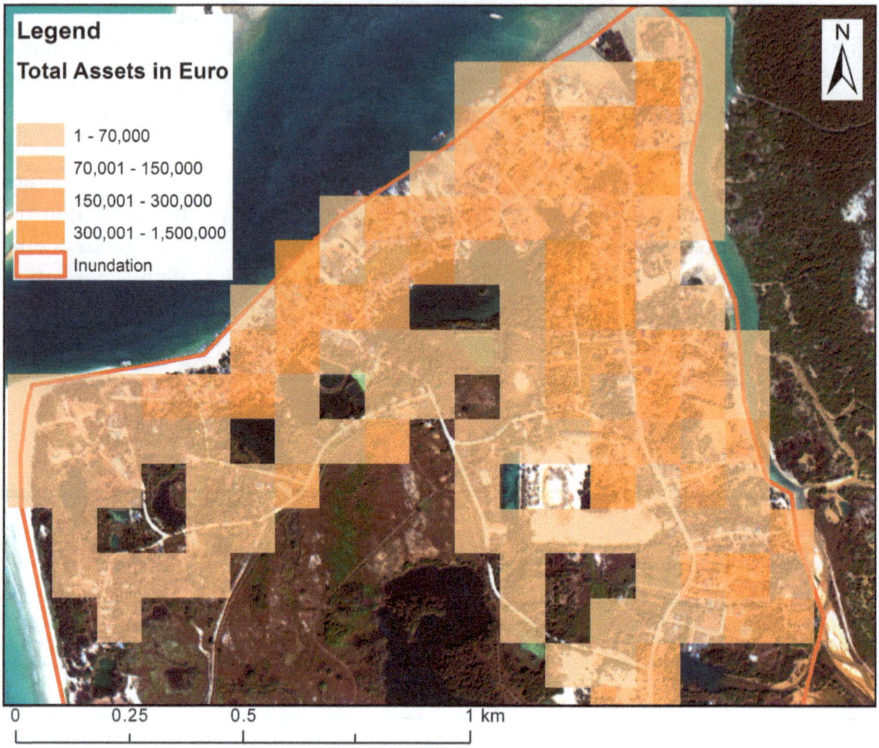

Map 8: Estimated exposure of household and company assets in Ban Nam Khem 2004 (Author's design, extrapolation based on TRAIT household and company survey data and automatic building extraction from MFC data)

The number of companies in Ban Nam Khem was determined by the field mapping. In communities in rural areas, more than one-tenth of the households run a small business inside their private homes. This type of company represents the majority of businesses in Ban Nam Khem, with only a limited number of companies, especially from the manufacturing or fishing business, having their own building available. The major concentration of assets can be found in the 'agriculture and fishing' industry, especially in the fishing business (cp. Table 34). This industry in Ban Nam Khem is highly heterogeneous because of its mixture of small-scale fishing and market-oriented large-scale fishing. In addition, several companies are back-linked to the fishing industry, especially from the manufacturing and service sector providing logistics, storage, cooling and other maritime supplies and services. The distribution of fish, aquaculture products and other seafood is organised by companies acting as middlemen and selling the products at the nearby markets (MASSMANN 2010:56; TRAIT Company Survey; SEGSCHNEIDER et al. 2008:3). Small-scale fisheries function mostly as self-employment and are based on the labour of the household members. The large-scale fishery companies own one or more offshore trawlers. The supplies needed for these companies consist only of consumables from Ban Nam Khem. Investment goods, such as boats, machineries and other equipment, are obtained from suppliers all over the province or from the Bangkok Metropolitan Region.

Table 34: Exposure of companies in Ban Nam Khem 2004 (Extrapolation based on TRAIT company survey data and field mapping data)

	Number	**Thereof located in inundation zone**	**Employees**	**Total assets in 1,000 €**	**Annual sales volume in 1,000 €**	**Annual supply volume in 1,000 €**	**Annual wages and salaries in 1,000 €**
Agriculture and fishing	63	100%	756	3,429.4	806,3	305.5	128.19
Manufacturing	14	100%	28	378.1	780.9	541.9	67.39
Retail	8	100%	183	1,014.6	1,011.3	186.9	134.91
Restaurants and hotels	17	100%	48	341.8	34.6	20.3	3.24
Services	27	100%	27	190.2	119.2	48.0	36.32

The retail, restaurant and hotel and service industries in Ban Nam Khem are almost exclusively made up of small companies run by the owner and his/her family. The majority of the customers come from the same village, and tourists are only occasionally counted among customers. Suppliers of operating goods or final products sold in the retail shops are mostly located in the nearby district capital of Takua Pa.

In terms of exposed economic elements, Tha Din Daeng in the area of Thai Muang is the smallest community and has only marginal economic values. Due to the small community size of less than 300 residents and the amount of assets per household, which is below the average of all the other three communities in total, only 400,000 € worth of assets for all households can be estimated. The households are, without exception, located in safe areas which were not inundated during the tsunami (cp. Table 35).

Table 35: Exposure of households in Thai Muang 2004 (Extrapolation based on TRAIT household survey data and building extraction from IKONOS data)

	Total	**Thereof located in inundation zone**
Number of persons	263	0%
Amount of assets in 1,000 €	373.1	0%
Amount of income per month in 1,000 €	21.5	0%
Amount of expenditures per month in 1,000 €	11.7	0%

Only a quarter of the total monthly income is generated within the village. One-tenth of all income of the area comes from companies in Ban Nam Khem or Khao Lak. One-third is generated in other surrounding non-tsunami-endangered areas. Especially obvious is the fact that one-third of all income is generated in other southern provinces, and it can thus be counted in combination with the 5 % income from social transfers as remittances. Similar to Ban Nam Khem, the majority of everyday purchases are made directly in the community in the local retail shops. Mostly, goods with medium or long-term demand are purchased in Khao Lak or the nearby district capital (TRAIT Household Survey).

As seen before, the economy in this community is relatively small and the majority of people work in the surrounding areas. Additionally, the companies in the area are, with only a few exceptions, self-employed businesses which are run by the owner with help from family members (cp. Table 36).

Table 36: Exposure of companies in Thai Muang 2004 (Extrapolation based on TRAIT company survey)

	Number	**Thereof located in inundation zone**	**Employees**	**Total assets in 1,000 €**	**Annual sales volume in 1,000 €**	**Annual supply volume in 1,000€**	**Annual wages and salaries in 1,000€**
Agriculture and fishing	7	0 %	18	1,695.9	160.1	28.8	24.9
Manufacturing	1	0 %	1	119.5	3.1	0.4	0.0
Retail	6	0 %	12	253.2	239.0	500.6	0.8
Restaurant	3	0 %	5	236.5	186.9	131.7	0.0
Services	1	0 %	1	252.2	43.9	10.8	0.0

Therefore, the regional linkages are also relatively small. The targeted market is almost exclusively within the village itself or the bordering villages. Also, in terms of linkages to suppliers, the companies are strongly focused on the local area. As a consequence, 60 % of the suppliers are within the same village. The remaining 40 % of suppliers are located in nearby Khao Lak or on Phuket Island.

In Khao Lak, 2,616 buildings in total were extracted from the remote sensing pictures, of which 723 can be counted as residential buildings. The large number of company buildings results from the structure of the hotel resorts consisting of one big lobby and several smaller bungalows. Around half of all the buildings are located in the zone that was inundated by the 2004 tsunami. Since the majority of the businesses in the region are tourism-oriented, the majority of companies (76 %) are located inside the inundation zone. The majority of the households live outside the 2004 inundation zone (cp. Table 37; Map 9). Applying the regression model above would lead to personal casualties of around 1,690 persons for the region of Khao Lak.

Table 37: Exposure of households in Khao Lak 2004 (Extrapolation based on TRAIT household survey data and automatic building extraction from MFC data)

	Total	**Thereof located in inundation zone**
Number of persons	9,741	37.0%
Amount of assets in 1,000 €	17,800.2	37.0%
Amount of income per month in 1,000 €	751.1	36.4%
Amount of expenditures per month in 1,000 €	351.9	36.2%

The population of Khao Lak draws its income mostly from the region in which they live. More than 80 % of the workforce is employed directly in Khao Lak. Another 7 % of the population stated that they are dependent on transfer payments either from the government or from family members living in other parts of the country. The rest of the population is employed in nearby cities. The payment flows for consumption issues from households are mostly targeted at shops within the community. Short and medium-term consumable goods are purchased within the community in nearly 90 % of the cases. Purchases of long-term consumable assets, such as household appliances, electronic goods, furniture and vehicles, are not only limited to the local area. Only one-tenth of these items are purchased in Khao Lak and another 40 % in the nearby district capitals of Takua Thung and Takua Pa. For these items, a strong connection to Phuket Island can also be found. Since the Phuket Province is the hub for the development of the southern provinces and also an international tourism hotspot, the island accounts for one-fourth of all purchases in long-term consumable goods in Khao Lak.

Map 9: *Estimated exposure of household and company assets in Khao Lak 2004 (Author's design, extrapolation based on TRAIT household and company survey data and automatic building extraction from MFC data)*

In the case of the companies, the linkages to Phuket Island are also very strong. Investment goods in particular are purchased from Phuket Island. Nearly one-third of the investment goods are bought there, while in Khao Lak, only one-fourth of the purchases are made. In terms of the value of the purchased goods, the picture changes. The proportion of investment goods purchased in the Bangkok Metropolitan Region is nearly half of all purchases made, while the share of the Khao Lak and Phuket Provinces is reduced to 15 % and 12 % respectively. The companies mostly obtain basic corporate services and agricultural products from suppliers in the region and its surrounding villages. High value-added machinery and equipment, especially for the larger businesses involved in the tourism industry, is purchased from Bangkok or the Phuket Province. A similar situation can be found for the economic interdependencies in the regular supply of working materials or consumables. While nearly 40 % of the purchases for supply are made in the area of Khao Lak, the value of these purchases represents only 2 % of the total purchases made from the companies. Purchases made in the Phuket Province account for 20 % of the number and for only 3 % of the total amount of purchases. This is a sign of the strong development of Khao Lak as a special tourism hotspot, which is not so much linked to the surrounding area any more, but is more integrated in the national processes of Thailand. The tourism sector in particular is formed by several international hotel chains offering large facilities in the Khao Lak area. In this sector, the number of employees ranges between two workers from the household working in in-kind service to more than 300 employees working in the approximately 15 larger hotels. The sales volume also shows a wide range, with a variation coefficient of nearly 200 %.

Table 38: Exposure of companies in Khao Lak in 2004 (Extrapolation based on TRAIT company survey data and field mapping data)

	Number	**Thereof located in inundation zone**	**Employees**	**Total assets in 1,000 €**	**Annual sales volume in 1,000 €**	**Annual supply volume in 1,000 €**	**Annual wages and salaries in 1,000 €**
Agriculture and fishing	28	46.4%	56	728.0	143.4	53.3	22.8
Manufacturing	22	54.5%	391	327.2	1,963.1	1,362.2	169.4
Retail	109	18.8%	257	2,745.3	4,115.1	760.5	548.9
Restaurants	66	38.0%	374	6,374.7	3,502.9	2,057.6	328.6
Services	11	41.0%	325	6,317.6	1,917.5	772.2	584.3
Hotels	58	93.3%	1,869	669,467.3	9,421.7	5,289.4	1,516.9

The hotels in particular are located inside the inundation zone of 2004. Nearly all hotels and resorts are built along the coastline within a distance of 500 metres from the beach front in order to benefit from the direct access to the beach. The other tourist-oriented facilities, such as restaurants or service industries, are located behind these hotels along the main road Phetkasem in the villages of La On, Bang Niang and Khuk Khak. The strong dependence on tourists as a major income source is mainly found in companies located near the hotels, and therefore also located within the inundation zone. The median value for the sales volume for tourists from abroad or domestic tourists is around 95 % for the service industries and 100 % for the restaurants. In the retail sector, the distribution of sales volume between the local population and tourists is a little more even. While the retail shops along the roads leading to the beach resorts depend almost completely on tourists, shops along the main road and in the residential areas are only frequented by locals. In the case of the 'agriculture and fishing sector', no direct sales to tourists are reported. Nevertheless, these companies sell their products to agents acting as middlemen between the central market and the producers. Within the economic cycle, these companies are also supposed to be connected to the tourism-dependent industries, which could be seen in the analysis of the input-output linkages. The manufacturing sector is also dependent on the tourism-related industries, since the companies involved in this sector in Khao Lak are manufacturers of food products or for building companies which are linked to the hotels with contracts.

In contrast to Khao Lak, the municipality of Patong Beach has a much more urban structure and is more densely populated, with around 20,000 inhabitants in the area in 2004. The city is located at the end of the Patong Bay inside a valley. The mountains around the valley climb up very steeply to heights of around 100 metres, where there is still settlement area. Therefore, only a minority of the city area was inundated by the tsunami in 2004. The economy of Patong Beach is almost completely based on tourism, i.e. hotels and restaurants, and tourism-related businesses, i.e. the service industry and retail. Only some businesses can be counted in the manufacturing sector. These businesses are mostly manufacturers of textiles which are sold in their own shops, but some construction companies are also located within the city limits. Parallel to the situation in Khao Lak, these businesses prefer the proximity to the main tourist attraction of the city, namely the beach front. During the tsunami of 2004, the water inundated an area of around 2.4 km^2, reaching around one kilometre inland in some places that were not protected by buildings. In other more densely built-up areas, the water reached around 400 m inland, mostly channelled through the streets running straight into the hinterland. Since the inundation area is mostly used for business purposes, only a few households are located in this area, especially in the upper levels of combined business and residential buildings (cp. Map 10). By using the building classification, field mapping and the regression model, it can be estimated that around 3,500 persons live in this area, and that consequently, 22 % of the total household assets of the population are located in the inundation zone (cp. Table 39).

Table 39: Exposure of households in Patong Beach 2004 (Extrapolation based on TRAIT household survey data and automatic building extraction from MFC data)

	Total	**Thereof located in inundation zone**
Number of persons	16,423	21.7%
Amount of assets in million €	55.41	22.4%
Amount of income per month in 1,000 €	3,353.8	22.1%
Amount of expenditures per month in 1,000 €	1,194.5	23.9%

Household assets in Patong Beach are the highest of all four study communities with a mean value of more than 11,000 €, compared to 3,800 € in Thai Muang, 9,600 € in Ban Nam Khem and 6,100 € in Khao Lak. This is despite the fact that in Patong Beach, the lowest rate of households can be found that are the owners of their property.

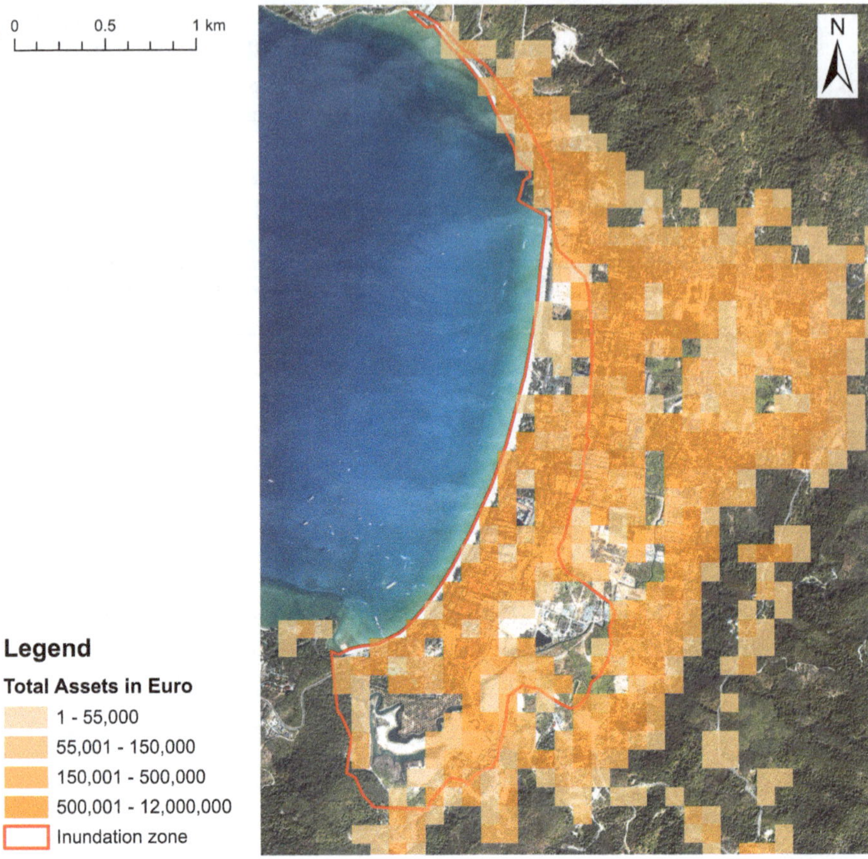

Map 10: Estimated exposure of household and company assets in Patong Beach 2004 (Author's own design, based on own calculation)

While in all communities in the Phang-Nga Province, around 84 % of the households are the owners of their property, in Patong Beach, only 46 % state that they own the apartment or house in which they live. The average household income in Patong Beach is around 830 €, and thus nearly twice as high as in Ban Nam Khem and Thai Muang. The variation coefficient also shows a lower range of income for Patong Beach than in Khao Lak (560 €) and in Thai Muang, which indicates a more equal distribution of the income than in the other communities. More than 94 % of the income of the population of Patong Beach is directly generated in the city. In contrast to the other communities, only 2.5 % of the population is dependent on transfer payments from governmental agencies or family members living outside the household. 37 % of the income in Patong

Beach is generated in the service sector indirectly dependent on tourism expenditures, while another 30 % is generated in the retail sector, which also belongs to the extended tourism-related industries. The hotel and restaurant businesses, whose employees account for another 20 % of the total income of the households in the municipality, are directly linked to the tourism industry. Due to the size of the city and the external cash inflows from the tourists, the city offers a lot of retail shops. As a result, 65 % of the short and medium-term expenditures of the households occur in Patong Beach, and another third of these expenditures is spent in the capital of the Phuket Province. In terms of the long-term consumable assets, such as electronics, vehicles and furniture, 70 % of the purchases are also made in Phuket Town and another 25 % are spent directly in Patong Beach.

Looking at the company income side reveals the strong dependency of the city on the tourism business. Nearly all companies depend more than 60 % on the tourists as direct customers. The variation coefficient also shows that the standard deviation of the income that is generated by tourists is smaller than the mean value, indicating that almost all shops are dependent on external flow of income from tourism. External linkages of the companies in terms of sales connections are mostly targeted at the capital Bangkok or other countries, especially in the case of the hotels operating together with tour companies in Europe or Australia. One-third of the companies in the city can satisfy their demand in supplies from other companies in Patong Beach. Another 35 % of the companies' demand comes from suppliers in the capital Phuket Town. One-third of the supply is also transported from the capital of the Kingdom of Thailand to Patong Beach. In terms of long-term investment goods, the suppliers are also located within the province of Phuket. 12 % of the suppliers are located in Patong Beach and another 65 % are located in the Phuket Province. Despite this distribution of suppliers for the companies in Patong Beach resulting from the share of the absolute value of the goods, it can still be seen that the more specialised and more highly valued goods need to be imported from the capital Bangkok. From Bangkok, 22 % of the total amount of investment goods is purchased, while in absolute terms, this represents not more than 10 % of the total transactions. On the other hand, the purchases from Phuket Island represent 85 % of the total transactions, but only account for 50 % of the monetary volume. Businesses in particular which are branches of a domestic or international company are strongly linked to their mother company and often do not depend on local suppliers. In the sample, 15 % of the companies stated that they were a part of a multi-branch company with the head office either on the island of Phuket or, in some cases, a domestic or multinational company. As in Khao Lak, 3 % of the companies stated having their head office in the capital Bangkok, and another 1 % are part of a multinational company. In reality, this share should be higher due to the limitations of a face-to-face interview-based survey, which often results in a lower response from larger companies due to the restrictions on data.

The highest concentration of assets in the company sector can be found in the hotel businesses, which account for nearly 95 % of the total assets in the area. Sales and supply are also dominated by this sector (cp. Table 40). Since Patong Beach provides nearly 20,000 jobs, many people commute to the community every day from the surrounding district of Kathu or from other districts of Phuket Island. It can be estimated that in Patong Beach, the workforce is around 10,000 persons, meaning that nearly half of the employees are from outside the city. As a result, an interruption of business activities in the community of Patong due to the impact of a natural hazard would have consequences for the population in other parts of the province, and would therefore spread out to other parts of the Phuket Province.

Table 40: Exposure of companies in Patong Beach 2004 (Own calculations based on TRAIT company survey)

	Number	**Thereof located in inundation zone**	**Employees**	**Total assets in million €**	**Annual sales volume in million €**	**Annual supply volume in million €**	**Annual wages and salaries in million €**
Manufacturing	27	0.0%	144	0.03	0.13	0.08	0.01
Retail	459	40.%	1,153	10.9	4.34	0.80	0.58
Restaurant	436	32.3%	3,018	163.1	21.34	12.53	2.00
Services	388	33.8%	1,358	14.8	7.22	2.91	2.20
Hotels	194	30.4%	13,626	96,889.2	795.92	446.83	128.14

7.2.3 Influencing factors on the vulnerability of an economic system

The fishing community of Ban Nam Khem experienced a rapid growth during the last century. The community started 80 years ago with only a few families, and developed into a community with nearly 5,000 inhabitants prior to the tsunami. In the beginning, the area was used for small-scale farming until the Japanese troops started to build piers. Later, a canal to the east of the community was dug to allow fishing boats to go further inland. In 1956, tin deposits were found, which led to the establishment of two tin mining companies in the area and an increasing demand in labour. In the following years, the community grew rapidly due to the migration of people from other provinces as well as from nearby Burma. In the 1990s, the tin deposits were nearly exhausted and, at the same time, the tin prices on the world market dropped (THULER 2007:14).

As a consequence, the population turned to commercial fishing or to working in the growing tourism business in nearby Khao Lak, which can be seen in the strong concentration of employment in the fishing and the hotel and restaurant sectors measured by the location quotient[3] (cp. Table 41). Prior to the tsunami, one-third of the total population worked directly in the agriculture and fishing sector, but mostly in small-scale businesses, since one-third of workers were self-employed (PHAPHAVASIT et al. 2009). Two-thirds of people working in this business do not have a working contract, which puts the employment in this business in a risky position without social security or other job protection. Despite this precarious job situation, the income level per person of 410 € in the fishing sector is above the monthly average of the Kingdom of Thailand. While the majority of sales in the large-scale fishing sector are exported, small-scale fishers also generate high incomes in this region. This is caused by the strong linkage and proximity of the community to the nearby tourism hotspot of Khao Lak. The majority of the sales from small-scale fishing go via middlemen to the nearby hotels and resorts in the area (INTERVIEW KHAO LAK HOTEL ASSOCIATION). Another important income source for the population in the community is the 'hotel and restaurant' and the 'retail' sectors. The majority of people involved in these business sectors are employed in the nearby Khao Lak area. Those commuting to Khao Lak are mostly employed as labourers, of which only half are employed with an official contract. Additionally, the income level of this group is only half that of people involved in the fishery sector (cp. Table 41).

3 The location quotient measures the relative concentration of a characteristic (employment, sales volume etc.) of a specific business sector in a regional sub-area. It is calculated as the sub-area's share of an economic activity in a particular industry in relation to a reference area's economic activity in the same industry. A quotient of one indicates a similar share, above one indicates a stronger concentration and coefficients smaller than one indicate a lower concentration (SCHÄTZL 2000:64).

Table 41: Employment situation in 2004 (Own calculation based on TRAIT household survey and NSO 2008a; NSO 2007c; NSO 2007d)

	Location quotient for employment concentration in						
	Agriculture	**Fishing**	**Manu-facturing**	**Retail**	**Hotel and Restaurant**	**Service**	**Average monthly household income 2004 in €**
Ban Nam Khem	0.1	22.5	0.4	1.5	2.0	1.0	407_a
Thai Muang	1.1	5.4	0.4	0.9	2.6	0.6	358_a
Khao Lak	0.3	8.0	0.8	1.0	3.0	1.3	558_a
Patong Beach	0.2	0.0	0.5	1.6	4.2	1.4	827_b
Phang-Nga	1.4	2.0	0.2	1.0	1.3	1.1	358
Phuket	0.1	0.7	0.7	1.3	4.5	1.5	532
Thailand	34.0%	1.4%	22.7%	16.9%	7.1%	18.0%	315

Average monthly income: Subscripts in the same row not sharing the same letters are significantly different at $p<0.05$ in the two-sided t-test

While the local businesses in Ban Nam Khem are mostly orientated towards agriculture and fishing, the households are much more diversified in terms of their income sources. The Shannon Index indicates a high diversification of income compared to the other study areas with an index value of 10.1. The index does not correlate with the total amount of household income, which is mostly the case when using the number of income sources per household as an indicator for diversification. Furthermore, the diversification index for the 7 % of households in Ban Nam Khem which are below the poverty line is significantly lower than for the rest of the households. At 7 %, the poverty incidence in Ban Nam Khem is much higher than the average for Phang-Nga, which is around 0.6 % (UNDP 2007:117). In the community, poverty is most likely among people working in the construction or manufacturing sector, which has the lowest salaries of all sectors in the community. Despite the high poverty incidence for Ban Nam Khem, the community is not an outward migration area. The age distribution for Ban Nam Khem indicates that the area is highly similar to the whole area of southern Thailand (cp. Table 42).

Table 42: Age distribution in study area 2009 (Own calculations based on TRAIT household survey and NSO 2000 for southern Thailand)

	Southern Thailand*	**Ban Nam Khem**	**Thai Muang**	**Khao Lak**	**Patong Beach**
0-4	7.7%	9.3%	7.8%	10.9%	4.5%
5-9	7.9%	5.9%	8.7%	8.3%	5.0%
10-14	8.6%	7.8%	10.4%	5.6%	6.6%
15-19	8.5%	9.8%	8.3%	3.8%	7.7%
20-24	8.4%	9.3%	7.4%	7.1%	9.8%
25-29	8.6%	8.8%	6.1%	9.0%	11.3%
30-34	8.4%	8.8%	7.8%	11.3%	11.9%
35-39	8.3%	6.6%	11.7%	9.0%	11.1%
40-44	7.6%	7.4%	7.4%	7.1%	7.7%
45-49	6.4%	5.9%	4.8%	6.4%	7.1%
50-54	5.1%	5.4%	5.2%	4.9%	5.0%
55-59	4.0%	4.9%	5.2%	3.0%	2.6%
60-64	2.8%	3.2%	2.2%	4.9%	2.4%
65-69	2.6%	1.5%	2.6%	1.9%	2.4%
70-74	2.1%	2.9%	1.7%	2.3%	1.6%
75-79	1.5%	1.7%	1.7%	3.4%	2.6%
80-84	0.9%	0.2%	0.4%	0.8%	0.3%
85 and older	0.7%	0.5%	0.4%	0.4%	0.5%
Spearman correlation with southern Thailand	--	0.91*	0.87*	0.74*	0.86*

correlation significant 2-side 1% - n = 635 household members *Data for 2000

The population of working age in particular accounts for similar shares to the whole southern area, which shows that there is no outward migration from the area. The percentage of children aged between 5 and 14 years, as well as the percentage of older people, is reduced significantly due to personal casualties caused by the tsunami, which is more likely to affect children and senior citizens.

Traditionally, the Thai economy is strongly shaped by the large number of small and micro-enterprises. Over the whole country, up to 60 % of companies can be classified as micro-enterprises (1-9 employees), and another 30 % of the companies are small enterprises with 10-49 employees (MOI 2003). Particularly in the rural areas and in times of economic crisis, this form of business experiences a strong growth (WIBOONCHUTIKULA 2002:3). In Ban Nam Khem, nearly 90 % of the enterprises can be classified as micro-enterprises according to official classification. The majority of them consist only of the owner and family members supporting him/her without payment. In the service, restaurant and retail businesses in particular, this form of company is dominant for the economic system in the community. The 'agriculture and fishing industries' and the manufacturing industries differ from this pattern. The companies in the agricultural and fishing sector, which is dominated by fishing in this community, fall mainly in the category of small enterprises, since the boats need to be manned by up to fifteen people. In the back-linked manufacturing sector, one-third of companies can be classified as medium enterprises, which employ 50 to 300 persons. In terms of sales levels, the companies in Ban Nam Khem can only be classified as small or micro-enterprises, showing sales levels for 2007 of under 430,000 €. In 2003, prior to the tsunami and the economic crisis, the sales level of the companies interviewed was slightly higher, with around one-fifth of them being classified as small enterprises. In 2007, their share was 14 %, while the rest of the companies were classified as micro-enterprises (cp. Table 43).

Table 43: Classification of companies according to sales and number of employees (Own calculations based on TRAIT company survey)

Employees 2003					Sales in 1,000 Euro 2003				
	Ban Nam Khem	**Thai Muang**	**Khao Lak**	**Patong Beach**		**Ban Nam Khem**	**Thai Muang**	**Khao Lak**	**Patong Beach**
Micro (0-4)	89.7%	100.0%	87.9%	84.9%	Micro (0-22)	81.1%	92.9%	67.9%	78.7%
Small (5-19)	5.1%	0.0%	7.6%	7.6%	Small (22-428)	18.9%	7.1%	30.2%	15.5%
Medium (20-99)	5.1%	0.0%	0.0%	3.5%	Medium (428-2,997)	0.0%	0.0%	0.0%	3.9%
Large (more than 100)	0.0%	0.0%	4.5%	4.1%	Large (more than 2,997)	0.0%	0.0%	1.9%	1.9%

Employees 2007					Sales in 1,000 Euro 2007				
	Ban Nam Khem	**Thai Muang**	**Khao Lak**	**Patong Beach**		**Ban Nam Khem**	**Thai Muang**	**Khao Lak**	**Patong Beach**
Micro (0-4)	92.3%	100.0%	83.3%	79.7%	Micro (0-22)	86.1%	100.0%	68.5%	72.1%
Small (5-19)	5.1%	0.0%	12.1%	11.6%	Small (22-428)	13.9%	0.0%	29.6%	22.1%
Medium (20-99)	2.6%	0.0%	1.5%	4.7%	Medium (428-2,997)	0.0%	0.0%	1.9%	3.2%
Large (more than 100)	0.0%	0.0%	3.0%	4.1%	Large (more than 2,997)	0.0%	0.0%	0.0%	2.6%

Mean values for all indicators are significant different on 5% error-level among communities using T-Test – n = 199 for 2004 and 295 for 2007

In Ban Nam Khem, formal education does not play an important role, especially in micro-enterprises. In these companies, the owner has finished lower secondary school, while the employees have an education of around six years, equivalent to primary school. A large number of Burmese refugees in this position can also be found, who mostly stated that they do not have any formal education. In small and medium-sized businesses, the education level among the managing employees is significantly higher on average. In small enterprises, the manager or owner has finished upper secondary school, and in medium enterprises on average, the owner or manager holds a diploma from a vocational college or a university, or a college bachelor degree. For the workers, no significant difference within the company classes can be found. One-third of the companies in Ban Nam Khem hire additional workers to cover the seasonal changes. This group of workers only possesses minimal education, which is mostly not more than primary school, and sometimes even less. In Ban Nam Khem, the 'agricultural and fishing sector' and its back-linked manufacturing industries in particular experience seasonal changes in demand. Some retail companies - around 20 % - also report experiencing this seasonal change in demand. These companies are also linked to the 'agriculture and fishing' sector through supply or demand connections. These seasonal changes follow the general climate in Thailand, and especially the occurrence of the south-western monsoon, which brings heavy rainfall in the country from May to September, meaning that fishing and agricultural activities need to be reduced. Another important criterion for the resilience capability of a company is its age, indicating the experience of a company in dealing with external shocks. In Ban Nam Khem, nearly every second company interviewed was younger than five years, and therefore founded after the occurrence of the tsunami in 2004. The rest of the companies older than five years were, on average, between 13 and 20 years old, without any strong deviation. As seen in the previous chapter, the economy of Ban Nam Khem is strongly focused on itself regarding sales and supplies. The Shannon index applied to the customer groups of each company shows similar values for all of the branches, with variation coefficients ranging between 5 and 15 %. A significantly higher sales diversification can only be found in the manufacturing sector, which is more orientated towards the national and provincial market providing processed fish products. Another criterion that is often mentioned and is supposed to increase the resilience capability of companies is the access to the banking system in order to receive loans and credits. In Ban Nam Khem, more than one-third of the companies have debts at the local banks in Ban Nam Khem itself or the district capital of Takua Pa (cp. Table 44).

Table 44: Companies' access to banking 2007 (Own calculations based on TRAIT company survey)

	Amount of deposit in	Amount of credit in €
Ban Nam Khem	$1{,}114.7_a$	$4{,}281.3_a$
Thai Muang	64.3_b	742.6_b
Khao Lak	$50{,}157.0_c$	$160{,}412.2_c$
Patong Beach	$99{,}702.8_d$	$32{,}167.8_d$

Subscripts in the same row not sharing the same letters are significantly different at $p<0.05$ in the two sided t-test – n = 199

Households in Thai Muang make up the group with the least exposed elements in absolute terms. Compared to the other study areas, the income level and the access to banking, measured in savings and credits, is the lowest of all study communities by a significant margin (cp. Table 41; Table 48). Similarly to Ban Nam Khem, nearly a quarter of the total household income is generated in the agriculture and fishing business, and one-tenth of the income comes from tourism-related businesses located mostly in nearby Khao Lak. Parallel to Ban Nam Khem the location quotient for employment in agriculture, fishing or hotel and restaurant businesses indicates a concentration above the national average (cp. Table 41). Despite this the households are significantly more diversified in their income situation than households in the other study areas (cp. Table 48).

On the side of the companies, it is obvious that nearly all companies are small-scale micro-enterprises with less than four employees and a sales level lower than 21,000 € per year (cp. Table 43). As a consequence of the low sales level, these companies only have limited access to the banking system, since they show the lowest debt and credits of all study areas (cp. Table 44). In terms of experience of the companies' owners, measured by the age of the company, no significant difference can be seen from the other study areas, except Ban Nam Khem (cp. Table 45).

Table 45: Age of companies in study area (Own calculations based on TRAIT company survey)

	Age of companies in 2004
Ban Nam Khem	12.4_a
Thai Muang	4.8_b
Khao Lak	4.5_b
Patong Beach	4.7_b

Subscripts in the same row not sharing the same letters are significantly different at $p<0.05$ in the two-sided t-test - n = 199 for 2004 and 295 for 2009

In terms of seasonal changes, the companies in the area are prone to the weather conditions, especially in the agriculture and fishing business, with a quarter of the companies reporting seasonal influences in their business. This corresponds significantly to the situation in Patong Beach, where the companies are influenced slightly by the weather due to changes in tourist numbers. This has only a marginal impact on the local employment situation because of the low employment level of the local companies.

Khao Lak in the south of Ban Nam Khem shows the lowest correlation according to its age structure compared with the southern part of Thailand. The share of persons between 25 and 40 in particular exceeds the age distribution in the southern provinces (cp. Table 42). This indicates that the community has a dynamic economic setting which attracts inward migrants from other parts of Phang-Nga and the rest of the kingdom. Additionally, it was found that in the surrounding communities of Ban Nam Khem and Thai Muang, a quarter of the workforce of Ban Nam Khem and one-third of the workforce of Thai Muang is employed in the area of Khao Lak and commutes daily to the area. The reason for this is that the income level in Khao Lak greatly exceeds the average incomes in the province of Phang-Nga. While the income level in Ban Nam Khem and Thai Muang is around 450 € on average per household, the income level in Khao Lak is nearly 70 % higher. In the potentially tourism-oriented businesses of retail, hotel, restaurant, and services in particular, the income level is between 40 and 100 % higher. As a result, Khao Lak has the lowest poverty rate of all study communities, with 3 % of the households being identified as poor according to the poverty line of the UNDP (2007:117). According to the situation in Ban Nam Khem, these households show a significantly lower diversification index for their household income. The level of diversification for Khao Lak is nearly similar to the one in Ban Nam Khem, reaching an index value of 10.6.

At the beginning of the 1990s, the area of Khao Lak was dominated by agriculture, like the whole province of Phang-Nga. The fishing sector along with rubber, fruit and coconut plantations in particular, dominated the economy of the area. Nowadays employment is concentrated in the fishing and in the hotel and restaurant sector (cp. Table 41). Starting in the mid-1990s, Khao Lak began a rapid development to become one of the tourism hotspots of Thailand (CALGARO 2005:37pp). Starting with around 100 rooms in 1996 distributed over small bungalow hotels, several new major hotels were located in the area over the next eight years, providing more than 6,000 rooms by the end of 2004. The development started almost exclusively without the participation of external companies and relied mostly on local businesses, which currently still provide 70 % of all the 840 rooms now in the area. In 2000, international chains entered the region and now run the larger hotels. During this time, a German travel company became beware of the destination and supported local businessmen with loans and credits (INTERVIEW KHAO LAK ECOTOURISM). Nowadays, the region's tourism industry focuses strongly on the upper segment in the market. Nearly half of all the 19 hotels which are now in Khao Lak are grouped in the highest category, charging an average of more than 200 € per night. The other end, with prices lower than 15 €, is not well equipped, since only three hotels are listed in these two categories (cp. Table 46).

Table 46: Accommodation in the study areas 2007 (Author's own design, based on TAT 2008a; TAT 2008b)

	Category	1	2	3	4	5	Total
Patong Beach	Mean price	146 €	53 €	28 €	20 €	10 €	47 €
	Number	34	33	30	68	29	194
	Share	17.5%	17.0	15.5%	35.1%	14.9%	100%
Khao Lak	Mean price	205 €	65 €	31 €	14 €	13 €	115 €
	Number	9	3	4	1	2	19
	Share	47.4%	15.8%	21.0%	5.3%	10%	100%

Depending on the seasonal weather changes, the economic activities in the tourism sector are limited to the time from October to March when the hotels are fully booked. During the rest of the time, the occupancy rate does not reach more than 30 %. Despite this, only one-third of the tourist-related companies in the hotel, restaurant, service or retail sectors hired additional employees during this time (cp. Table 47).

Table 47: Seasonal changes in employment 2009 (Own calculations based on TRAIT company survey)

	Companies subject to seasonal changes	**Companies reducing the number of employees due to seasonal changes**
Ban Nam Khem	18%$_{a}$	21.7%$_{a}$
Thai Muang	26%$_{b}$	8.3%$_{b}$
Khao Lak	52%$_{c}$	31.0%$_{c}$
Patong Beach	23%$_{a}$	23.5%$_{d}$
Agriculture and fishing	42.9%$_{a}$	14.3%$_{a,b}$
Manufacturing	37.5%$_{a}$	25.0%$_{a,b,c}$
Retail	11.3%$_{a}$	8.1%$_{a}$
Restaurant	30.0%$_{a}$	40.0%$_{b,c}$
Service	27.1%$_{a}$	16.7%$_{a,b}$
Hotels	61.1%$_{a}$	61.1%$_{c}$

Subscripts in the same row not sharing the same letters are significantly different at $p<0.05$ in the two-sided t-test – n = 295

Due to the high training requirements for the staff working in international hotels, the number of employees is kept constant all over the year and the employees are occupied with maintenance or training. Letting staff go during the low season would be more expensive due to the strong regional disparities in Thailand which offer highly skilled workers easy options for finding another more highly paid job in another part of the country (INTERVIEW KHAO LAK HOTEL ASSOCIATION). Only low-skilled workers for restaurants or services are hired during the high season, which only leads to marginal effects in the job market. In general, the education level in the companies in Khao Lak is significantly higher than in Ban Nam Khem. In all branches, the average education of the employees involved in management as well as general workers is that of a completed upper secondary education. Despite this, the education level for management in the service sector and the hotels is slightly higher and shows an average equivalent of a bachelor degree. Furthermore, the workers in the service sector also have, on average, a bachelor degree, which is a result of the services, in some cases very specific, that are offered from this sector. Nearly half of the companies of this sector are tour agencies offering and operating their own tours, which requires good language skills. Several real estate agencies and constructional engineering companies are also located in the Khao Lak area. These companies also target the international market, offering long-stay apartments or permanent recreational accommodation. As in Ban Nam Khem, the population of businesses in the area is relatively young. Slightly more than 40 % of all businesses were founded after the tsunami of 2004, and are therefore not older than five years. The age of the rest of the interviewed companies ranges between 5 and 20 years, with a division between tourism-oriented and locally oriented businesses. While the companies in the restaurant, hotel and service sector were, on average, founded in the year 2000, the retail and manufacturing companies are, on average, ten years older. In her study on the vulnerability of Khao Lak, CALGARO (2005:45) found that the local companies suffer from a lack of capital due to limitations in accessing the capital market. This statement could not be verified by this survey, since the majority of companies already have debts and credits, and only a minority of the companies from the restaurant and retail sector do not have liabilities with their banks. These companies are exclusively micro-enterprises or companies that already have access to foreign capital. Similar results could be found for the micro-enterprises in terms of the penetration with non-life insurance. Traditionally, the insurance rate is relatively low in emerging countries in general. In Thailand, 1.4 % of the GDP was insured in 2003 (INTUWONGES 2010). Of all companies in Khao Lak, around one-third do not have any insurance against damage to company property from natural hazards. For the micro-enterprises, this share reaches nearly 40 % of all companies which are not insured for any circumstances, and therefore do not have the opportunity to transfer risk. Only for the small and medium-scale companies is the penetration rate with insurance significantly higher, and reaches 100 % in the case of medium-scale companies.

Patong Beach is located on the west coast of Phuket Island in the district of Kathu. Patong Beach is one of the most famous tourism destinations on the island of Phuket. It is estimated that around 90 % of the population of Phuket Island is directly or indirectly engaged in tourism-related businesses (ICHINOSAWA 2006). Around one-fifth of all the hotels and guest houses of Phuket Island are concentrated in Patong Beach. In total, the municipality hosted around 200 hotels offering around 10,000 hotel rooms in 2004 before the tsunami. Patong Beach experienced a rapid development as a tourism hotspot. In the 1980s, only a few shacks were located at the beach of the village without paved roads and electricity, and the area was only a destination for backpackers. Most of today's urban area was used as a coconut plantation. Today, the TAT promotes the city as a perfectly developed beach with a huge variety of leisure activities and an exciting night life (TAT 2009a). The majority of the businesses are directly linked to tourism: one-third of the total businesses come under the 'hotel and restaurant' category, one-third are retail businesses and another third are involved in the service sector. Agricultural and manufacturing businesses are virtually absent from the community of Patong. Also the employment situation follows this trend. Employment in the Agriculture, Fishing and Manufacturing sector is nearly nonexistent, while the highest concentration of employment can be observed in the tourism and service sectors (cp. Table 41). More than 70 % of the companies state that domestic or international tourists are one of their three most important customer groups (TRAIT Company Survey 2009). Despite this, a lower percentage of companies in Patong Beach state that their business is often subdued by the effects of the seasonal changes. In total, only one-fourth of all companies believe to have seasonal changes in their business activities, while in Khao Lak, more than half of all companies are influenced by the seasonal changes. The reason for this situation is the more developed tourism infrastructure on Phuket Island, which makes it an attractive destination, even in the off-season. While the area of Khao Lak is oriented towards a recreational and nature-based tourism, with guests staying for two weeks on average (INTERVIEW KHAO LAK HOTEL ASSOCIATION), Patong Beach is oriented towards another market. While in Khao Lak, more than 60 % of the hotels are located in the upper two price segments, this is only the case in Patong Beach for one-third of the hotels. The majority of the hotels target the lower price segments (cp. Table 46). This clearly indicates that Patong Beach is strongly focused on the price-conscious mass tourism, and no strategic changes have occurred in the past decade (VORLAUFER 2005:16). Besides the nature and the beaches of the island, various large shopping malls, several amusement parks and indoor recreational facilities have been developed on the island to attract tourists in the rainy season as well. As a result, the changes in the number of employees over the seasons are slightly smaller than in Khao Lak. The size of the companies also differs from Khao Lak. While in Khao Lak, it is mainly micro-enterprises and small-enterprises that dominate the local economy, in Patong Beach, especially in the hotel sector, medium and large enterprises are in the majority. In the service, retail and restaurant sectors, small enterprises are also dominant, rather than just micro-enterprises. As a consequence of these differences in the size structure of the companies, the mean age of the businesses is slightly higher than in the other study areas. In general, one-third of the companies in Patong Beach are younger than five years, and were therefore founded after the tsunami. A special pattern can be seen for bigger companies, i.e. medium and large enterprises, which have, on average, been in business for around 12 years, while small and micro-enterprises are around seven years old. Another result of the shift towards companies with higher sales volume is the better access to credits and loans. Medium and large companies in particular almost all have bank obligations, while for small and micro-enterprises, only 40 % report borrowing money from the bank. Also related to the size of the company is the possibility of being a multi-branch company. 14 % of all companies interviewed in Patong Beach are multi-branch companies, with their head office either in Phuket Town or in the capital Bangkok. Additionally, every tenth company has foreign investors holding shares in the company. The demand for better educated staff is also following the trend to companies that show higher sales volumes and higher numbers of employees. In the hotel sector in particular, the average management employee needs a bachelor degree or higher, while the workers are also mostly educated to the level of upper secondary school or have a diploma from a vocational college. This education level is also spread among the other business sectors. Only among the working staff in restaurants and retail stores without management duties is the mean education on the level of lower secondary school, which might also be a result of the employment of unskilled migrant workers from other parts of the country or nearby Myanmar. This trend of inward migration towards Patong Beach can also be clarified by looking at the age distribution for the population living in the city. The age groups in the working range, i.e. 29-49 years, are clearly over-represented compared to the rest of the southern provinces (cp. Table 42). Furthermore, younger people under 19 and older citizens are strongly under-represented in Patong Beach. Besides the inward migration, the reason for this distribution might also be a suburbanisation process, with the result that people with higher incomes tend to move outside the city to the surrounding district of Kathu. Corresponding to the findings of the company survey, the households in Patong Beach have the highest education of all households in the observed study communities for the population of working age, with 9.3 years on average, which is almost one year more than in Ban Nam Khem. Resulting from the high education level and the strong concentration of jobs in the city, the income level for Patong Beach reaches more than 800 €, which is nearly double the value of that of Ban Nam Khem, and exceeds the average for the Kingdom of Thailand. More than one-fifth of the total income of the households is sourced in the tourism sector in the narrow sense, i.e. the hotels and restaurants. Since the economy in the city is almost completely oriented towards the tourism business, it can be assumed that the share of income generated in tourism businesses in the wider sense, i.e. tourism-oriented retail and services, is much higher (cp. Table 48). This results in a relatively low Shannon index for income diversification. In terms of the safety of the workplaces, no significant differences result from the special situation in Patong Beach. The percentage of employees who have an official contract with their employer is not significantly higher than in Khao Lak. The percentage of entrepreneurs among the population is also on a similar level to that of the other study areas.

Table 48: Household indicators for employment 2004 (Own calculations based on TRAIT household survey)

	Shannon index for income diversification	**Self-employed household members**	**Members with contract**	**Income from agriculture and fishing**	**Income from tourism**	**Mean education of working household members in years**	**Savings in €**	**Debt in €**
Ban Nam Khem	10.1_a	$26.8\%_a$	$7.2\%_a$	$23\%_a$	$11\%_a$	10.2_a	140_a	$1{,}609_a$
Thai Muang	10.3_b	$21.4\%_b$	$16.1\%_b$	$25\%_a$	$11\%_b$	10.1_a	130_a	451_a
Khao Lak	9.9_c	$36.7\%_c$	$11.8\%_c$	$12\%_b$	$16\%_c$	10.8_a	289_a	$2{,}045_a$
Patong Beach	8.2_d	$23.9\%_d$	$11.9\%_d$	$5\%_b$	$21\%_d$	11.0_a	179_a	$1{,}872_a$

Subscrips in the same row not sharing the same letters are significantly different at $p<0.05$ in the two-sided t-test – n = 171

Similar to the situation among the companies, most of the households do not have any insurance. In Ban Nam Khem and Khao Lak, around 55 % of the interviewed households did not have any insurance, despite the governmental health insurance. In Patong Beach, nearly 70 % of the households are insured. Life insurance is common for 50 % of the households, but only a minority of the households have an insurance policy which covers damages to buildings or household property from natural hazards. This is only the case for 1 % of the households interviewed.

7.2.4 Value chain linkages determining vulnerability

As discussed previously, vulnerability and recovery of companies is dependent on their capabilities to mitigate both direct and indirect influences of natural hazards. While in the case of a tsunami, the direct impacts depend on the location of the company, indirect effects depend on the status of the company within the economic cycle. This status determines the degree of affectedness in terms of the interruption of supply and sales, and the possibilities to shift to new business partners or customer groups. These interrelations of the business sectors can be identified with an input-output table analysis.

These indirect effects of a natural hazard on companies or households are dependent on their linkages and dependencies on companies or customer groups within the directly affected area. Another criterion for the spreading of indirect effects is the size and the economic relevance of the area for the country's economy. Impacts clearly restricted to a limited area, causing destruction of companies in these particular areas and the resulting reduction of output, can easily be compensated for by goods from other areas or imports. Companies losing their sales areas might also be able to switch their distribution areas to other parts of the country. Companies only serving a limited area or offering services that can only be performed directly to the customer, and are thus attached to a certain place, might suffer from the highest indirect effects, since they cannot easily switch to another distribution area. This refers especially to retail shops, various services and the place-based tourism industry. In terms of the I-O analysis, this means that industries located at the beginning of the economic cycle and which have more back-linkages to other companies are less likely to suffer from indirect effects caused by shortages in supply than industries located more at the end. In the case of industries targeting the final demand, the degree of localisation of the sales activities and the possibility to substitute suppliers is a key element in their vulnerability. The 'agriculture and fishing' sector and the 'manufacturing' sector are particularly integrated into the economic circle at an early stage, and mostly depend on sales to other sectors (cp. Table 49).

Table 49: Backward and forward linkages (Own calculation based on NESDB 2006; NESDB 2009)

	Backward linkages		**Forward linkages**		**Contribution to GDP**	
	in %	**Rank**	**in %**	**Rank**	**in %**	**Rank**
Agriculture and fishing	102	5	299	1	10.3%	4
Manufacturing	195	1	210	2	44.0%	1
Retail	46	6	0	6	14.7%	3
Restaurant and drinking places	157	2	20	5	0.9%	6
Hotel and lodging place	154	3	35	4	4.0%	5
Services	108	4	59	3	26.2%	2

Two other outstanding sectors can be found in the tourism sector with the hotel and restaurant sector, which is strongly linked to the manufacturing as well as to the agriculture and fishing sector. This sector holds a strong position as a key element in the circle of the economic system, although its contribution to the GDP is not so highly ranked and accounts for not more than 5 % of the national GDP. Forward linkages above the average can only be found in the agriculture and fishing and the manufacturing industries.

In the case of an external shock causing direct damage to production facilities or interfering in the logistics process, the interruption can cause indirect effects in all backward-linked businesses down to the households, which might suffer from losses of income. This is only the case for manufacturing and agriculture and fishing with a small distribution area. In the case of the study area in the Phang-Nga and Phuket Provinces, the manufacturing sector is limited to small facilities that mainly serve the nearby areas. Therefore, the results from the survey concerning the companies that were only indirectly affected indicate that these companies mostly suffered from declines in sales due to the effects on the back-linked industries. Nearly all of the companies involved in agriculture and fishing and two-thirds of the manufacturing companies interviewed confirmed a negative effect on their business due to reductions in demand. On the other hand, problems arising from the interruption of supplies were of minor importance. Here, only one-third of the companies not directly affected stated experiencing negative influences on their business due to this factor. The results from the back-linked industries indicate an opposing picture. The non-directly affected companies, especially the retail business, did not experience problems concerning the supply situation. Here, only 15 % reported issues regarding supply in the aftermath of the tsunami. In the industries involving restaurants, hotels and services, around half of the companies not directly affected reported having problems with supply. The reason for this was that these industries could establish new connections for their supply needs. On the other hand, a clear majority of around three-fourth experienced problems linked to sales resulting from the reduction of income or the absence of tourists. The 'service' sector in particular, i.e. restaurants, hotels and services, target private consumption or the 'special exports' which, in this case, mainly consists of consumption by foreign tourists in Thailand[4]. While the restaurant sector mainly depends on national private consumption, the hotel sector generates the majority of its value creation from foreign tourists (cp. Table 50). Around one-third of the value addition of the 'agriculture and fishing' and 'manufacturing' sectors depends on private consumption, while the 'special exports' only play a minor role. The manufacturing sector shows a particularly high dependency on 'exports'.

Table 50: Final demands in the economic system (Own calculations based on TRAIT company survey, NESDB 2006)

	Private consumption	**Exports**	**Special exports**
Agriculture and fishing	36.4%	5.0%	0.8%
Manufacturing	27.2%	38.4%	2.5%
Retail	0.2%	0.0%	0.0%
Restaurant	76.4%	0.0%	14.6%
Hotel	30.7%	0.0%	72.0%
Services	33.1%	0.0%	12.1%

The share of value addition from the retail and wholesale sector targeted at the final demand is remarkably low. This is due to the special nature of input-output analysis, which assigns retail and wholesale margins directly to the producing sector and not the retail sector itself. Only the actual value addition produced by the retail and wholesale sector in contact with the final consumer is reported here. A separate share for retail and wholesale trade margins for 'special exports' is not stated (NESDB 2006). Despite this, the margin generated with the consumers is shown separately, with 45 % of the margin being created

4 Remark: The balance of payments also included expenditures of Thai nationals abroad with the consequence that the balance might exceed 100 %

in the retail business with the final demand and 55 % being generated in the wholesale trade between companies or middlemen and retailers. This ratio is a result of the aggregation of the trading sector. As a result, the majority of the companies stated that they had waited for the return of these customers rather than establishing new customer groups (TRAIT Company Survey).

Analysing the integration of businesses into the economic cycle on the community level, only limited significant differences can be found. Therefore, the mean values for the I-O key figures of all companies in the study areas were calculated. In terms of backward linkages, no significant differences among the communities can be seen, despite the slightly lower mean value in the communities of Ban Nam Khem and Thai Muang. This effect is caused by the 'agriculture and fishing' industries, which are situated at the beginning of the value chain. Despite this, no significant differences can be found due to the large number of small retail businesses in the community. The high share of retail stores in Ban Nam Khem and the service-oriented businesses in Khao Lak and Patong Beach lead to relatively low values in terms of forward linkages. Here, Thai Muang, with its 'agriculture and fishing' businesses, shows a significantly higher integration in the economic cycle. Also, in terms of the two categories describing the orientation of the business sector towards final demand, i.e. private consumption and special exports, no significant differences between the communities can be found. Nevertheless, Patong Beach and Khao Lak show mean values that are around ten percentage points higher than in Ban Nam Khem and Thai Muang. This non-significant difference results from the service-oriented business structure in both Patong Beach and Khao Lak. Since the results among the communities only vary non-significantly, further analysis on the micro-level is necessary (cp. Table 51).

Table 51: Integration of companies in economic cycle 2004 (Own calculations based on NESDB 2006)

	Backward linkages	**Forward linkages**	**Private consumption**	**Special exports**
Ban Nam Khem	88_a	38_a	4_a	22_a
Thai Muang	80_a	122_b	2_a	22_a
Khao Lak	102_a	$61_{a,b}$	10_a	30_a
Patong Beach	107_a	40_a	13_a	31_a

Subscripts in the same row not sharing the same letters are significantly different at $p<0.05$ in the two-sided t-test

The analysis of the input-output table in combination with the survey data showed that the indirect effects of companies differ according to their position in the economic cycle: while larger companies at the beginning of the process might be able to switch to other markets, companies at the end of the chain depend particularly strongly on their customer base. The repercussions resulting from the employees being affected influence all businesses equally.

7.3 Effects of the 2004 tsunami in the study areas

7.3.1 Modelling the economic impact on the micro-level

In order to extrapolate the economic impact of the 2004 tsunami to all households and companies, two models were developed which should allow an evaluation of the amount of damage on the micro-level. These models reproducing the damage in relation to the physical sensitivity are part of the complete set of vulnerability models for households and companies which will be introduced in their totality in the following chapter.

Households

To represent the loss suffered by a household during the tsunami of 2004, household possessions consisting of property value and long-term consumable assets were chosen. The degree of the tsunami impact on the exposed household assets, and thus the sensitivity dimension, is represented by the location of the household. This factor is determined by the height of the household above sea level, the distance from the coastline and an indicator representing the degree of damage to the building derived from the statement as to whether a total rebuilding was necessary after the wave impact (cp. Appendix - Table 70). Both household damage-determining factors show a strong significant influence on this factor. While a greater amount of assets increases the damage, the location factor reduces the amount of damage. The strongest influence comes from the location factor, since it defines the degree of total household property affected, while the factor representing the exposed household's assets stands for the maximum possible damage. The explanatory power of these factors for the damage of the tsunami can be deemed 'weak', since the corrected coefficient of determination shows a value of 0.27 (CHIN 2009). If the fact is taken into account that only four items are taken to represent the total value of exposed household assets, this 'weak' R^2 can be accepted (cp. Table 52).

Table 52: Factors influencing amount of direct damage in households (Own calculations, based on TRAIT household survey)

	Path-coefficient	f^2	VIF	corr. R^2
Assets	0.25*	0.09	1.00	0.27
Location	-0.47*	0.30	1.00	
*significant min 5 % error- TRAIT Household Survey – Durbin-Watson: 2.07 – n = 171				

Companies

To represent the direct effects of the business sector, property value and productive assets were chosen. Floating assets could not be integrated into the model due to the diversified sample of different business sectors. Vehicles, as a part of the company assets, were not integrated due to a non-significant relation with the location of the company. In order to reproduce the physical tsunami damage caused by the wave impact, the pre-tsunami values available in the company in combination with the location of the company were regressed together on the dependent factor of actual tsunami damage to the property and production assets. Combining the influence of these two regressors; a sufficient corrected R^2 value of 0.52 can be observed. However, the power of influence represented by the f^2 value differs between the two independent factors. The location factor, set up by the indicators 'height above sea level' and 'distance from the coastline' for each sampled company, reaches an influential power which can be deemed 'weak', since the f^2 value reaches not more than 0.03. The outer loadings on the location factor represent the different influences of each indicator. Since the 'height above sea level' in the four study communities for the affected companies varies between zero and ten metres, and the 'distance from the coastline' varies between 10 and 1.5 km, the 'height' has a loading of 0.3 from this proportion, and the 'distance' reaches a loading of 0.8 (cp. Table 53).

Table 53: Factors influencing amount of direct damage in companies (Own calculations based on TRAIT household survey)

	Path-coefficient	f^2	VIF	corr. R^2
Assets	0.69*	0.99	1.01	0.52
Location	-0.11*	0.03	1.01	
*significant min 5% error - TRAIT Household Survey - Durbin-Watson: 2.08 - n = 199				

The R^2 value as an overall measure of fit for the amount of damage indicates a good fit of the chosen indicators with a value of 0.52. The explanatory power of the pre-tsunami amount of assets reaches an f^2 value of 0.99, which dominates the factor of 'damage to assets' per company. The reasons for this strong domination in terms of the explanatory power are the strong disparities between the companies in the region (cp. Chapter 7.2.3). Investigating the pre-tsunami assets in the different study regions shows the strong variation of the level of assets available in the companies. Here, the range differs between around 13,000 € in Thai Muang and more than 100 million € in Patong Beach on average. As a result, the stronger representation of medium and large enterprises in Patong Beach expresses itself in a variation coefficient of 9.6, whereas in Khao Lak and Ban Nam Khem, variation coefficients of 2.3 and 3.4 respectively can be observed.

7.3.2 Evaluation of the economic effects of the 2004 tsunami on the micro level

Due to its location, Ban Nam Khem was hit hardest by the tsunami waves. On average, the households suffered a loss of more than 12,000 € for the affected households (cp. Table 54).

Table 54: Direct impact on households by the 2004 tsunami (Own calculations based on TRAIT household survey)

	Total damage to assets in €	**Total time to recover in months**
Ban Nam Khem	$12{,}301_{a}$	9.8_{a}
Thai Muang	$1{,}707_{b}$	$6.4_{a,b}$
Khao Lak	$9{,}031_{b}$	5.3_{b}
Patong Beach	$5{,}789_{b}$	$3.6_{b,c}$

Subscripts in the same row not sharing the same letters are significantly different at $p<0.05$ in the two-sided t-test - n = 171

More than 90 % of the households reported damage to their house, totalling an average of around 4,300 €, which, in most of the cases, represents a total destruction of the building, since the reported damage was equivalent to the value of the property. Hand in hand with the complete devastation of the building, the interior of the households was also damaged, which accounted for around one-tenth of the total damage suffered by the households, or around 2,000 €. In general, nearly all households in Ban Nam Khem stated having suffered physical damage to their household property. 40 % of the households also experienced personal casualties among family members who died because of the disaster, and another 40 % of the households stated having had family members injured by the waves. In the time following the disaster, indirect effects resulting from the loss of jobs or the destruction of infrastructure particularly affected the households in their recovery activities (cp. Table 55).

Table 55: Indirect impact on households (Own calculations based on TRAIT household survey)

	Months with loss of income	**Months with effects from interrupted infrastructure on household**
Ban Nam Khem	14.5_{a}	0.8_{a}
Khao Lak	$4.3_{b,c}$	0.3_{b}
Thai Muang	$9.8_{a,b}$	0.5_{b}
Patong Beach	4.2_{c}	0.8_{a}

Subscripts in the same row not sharing the same letters are significantly different at $p<0.05$ in the two-sided t-test – n = 171

Household members working in the tourism-related hotel and restaurant sector experienced a particularly severe loss of income and/or unemployment for nearly one year. Persons employed in the agriculture and fishing sector still experienced reduced income for more than nine months, which is significantly above the average of 7.8 for reduced income for all persons in the workforce. Indirect effects resulting from the interruption of infrastructure played only a short-term role, since these problems were sorted out by infrastructure-providing companies and governmental agencies after two weeks (telephone network) or three months (road system). Among the tsunami-affected households in Ban Nam Khem, one-third of the households reported having had problems doing their jobs as a result of the damage to infrastructure. In terms of the rehabilitation process for their household, on average 60 % of the households interviewed experienced limitations resulting from the damaged infrastructure.

This regression model, as presented in Table 52, reproduces the amount of damage for each household in Ban Nam Khem, showing a mean value of nearly 11,000 € and a corrected R^2 value of 0.27 with significant interrelations on a two-sided 5 % error level. The average damage level assessed in the household survey reached around 12,300 €, meaning that the survey-assessed damage had been underestimated by the model by around 10 %, which can be accepted. In order to apply this model to the community of Ban Nam Khem, the buildings were classified using the field mapping results and a database of commercial and infrastructure use by the DDPM. Due to the physical exposure of the community, nearly all households were affected directly by the tsunami (cp. Map 11).

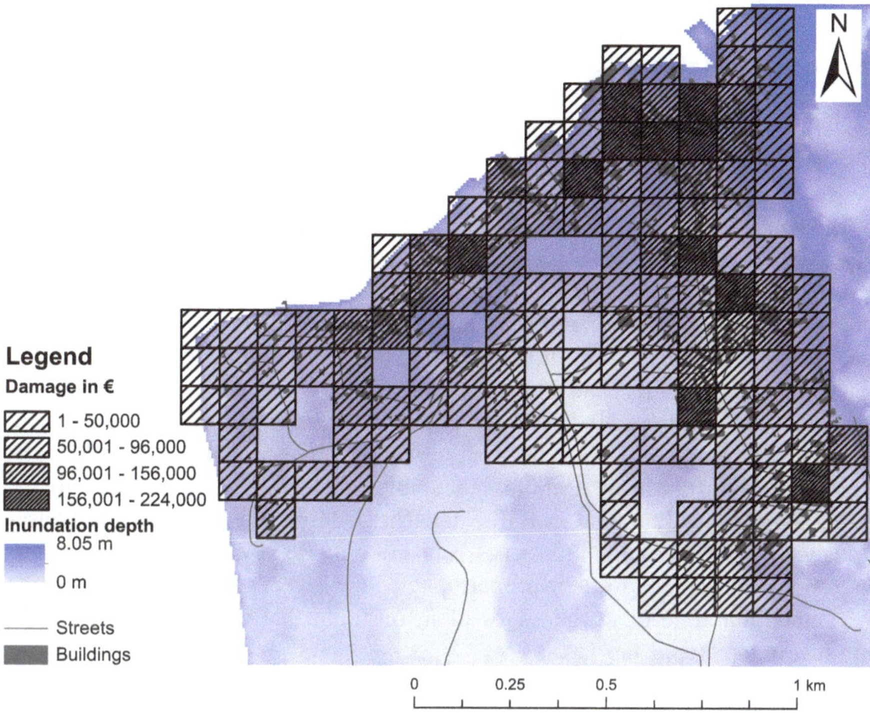

Map 11: *Inundation height and accumulated damage in Ban Nam Khem (Based on own calculation on TRAIT household and company survey and TRAIT Tsunami simulation data)*

The total damage for the 590 households comes to a figure of 6.2 million €. In addition to these direct effects, the households suffered from loss of income due to the impact on the business sector. On average, around five months of reduced income was the result of the tsunami for nearly all of the households in Ban Nam Khem. During this time, the income was reduced to the official poverty line of 30 € per person, since the government started its social security programme after the disaster (cp. Chapter 6.4). This leads to an accumulated loss of nearly the same amount as the direct damage to the households, with around 6.5 million € over the whole period of nearly 20 months. As a result, the expenditures of the households were reduced, causing negative effects for the companies. Parallel to the households, nearly all companies in the village were affected by the wave impact. On average, a company in Ban Nam Khem suffered damage to its property of nearly 14,000 €, standing against total company assets of nearly 60,000 €. For the approximately three hundred companies, this results in nearly 5 million € total damage to property and production assets. Indirect effects in terms of loss of sales for the businesses in Ban Nam Khem add up to 6.1 million €, since all businesses report a reduction of sales after the disaster without significant differences between the business sectors. On average, the companies restarted their business after around one year and continued their activities for one and a half years with average sales levels of around 50 % compared to the pre-disaster status.

In Khao Lak, south of Ban Nam Khem, the share of households affected physically by the tsunami waves was only around 36 %, since the majority of households were located outside the inundation zone, which mainly contained businesses. Therefore, the total accumulated damage of the affected households reached not more than 6.6 million €. Despite the fact that the physical damage to household property in Khao Lak was limited, the households had to carry the burden of the enormous indirect tsunami effects. Suffering from a reduced income for nearly 18 months with a significantly higher income level, the total indirect effects for the households accumulated to 19.5 million € due to the higher average income than in Ban Nam Khem. More than 60 % of the households were affected by the indirect effects, but only 36 % of the households suffered physical damage to their property, since they were mostly located outside the inundation zone or at the edges of it, meaning that they only experienced minimal impact (cp. Map 12).

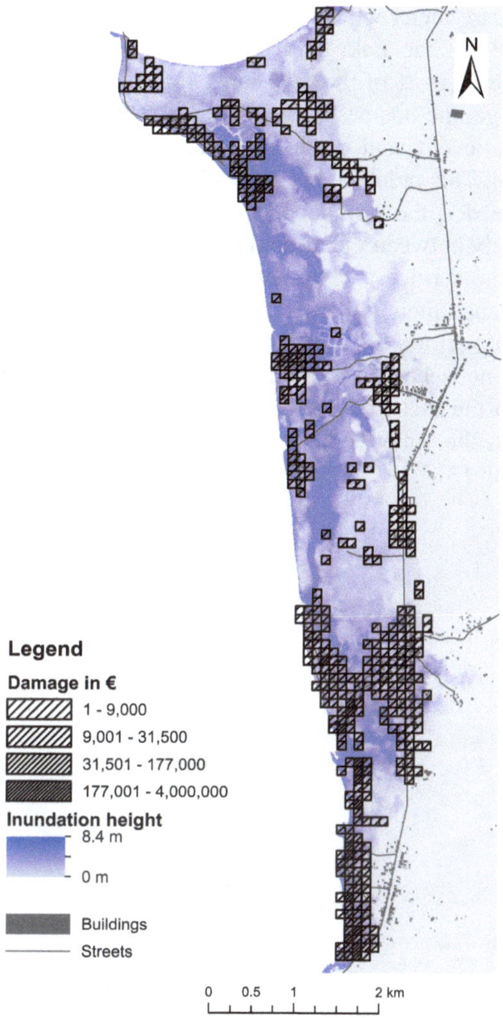

Map 12: Inundation height and accumulated damage in Khao Lak (Based on own calculation on TRAIT household and company survey and TRAIT Tsunami simulation data)

Hotspots for physical damage can especially be observed in the agglomerations of hotel complexes in the southern part of Khao Lak. Only along the main road in the village of Bang Niang could significant impact on residential buildings be observed. Comparing the amount of damage per square meter, it is obvious that the hotel business in the area dominated the total damage, with an average loss per square meter which is more than double the amount of the residential buildings. For Khao Lak, the calculated total reduction of sales accounted for 5.7 million € or 26 % of the total sales volume in the two years after the tsunami. On average, the businesses in Khao Lak started to work again around ten months after the event, and reached a total output of half of their pre-tsunami production level. This reduced output level continued for another year before the companies exceeded the pre-tsunami production level. Significant differences can be found only between the agriculture and fishing and manufacturing on the one hand and the service and tourism-oriented sectors on the other hand. The first sectors reached their full production level nearly half a year before the service and tourism sectors.

Damage to households in Patong Beach could only be observed in limited amounts due to the fact that the settlement areas in the municipality are more oriented towards the hinterland, while the coastal area is dominated by commercial use (cp. Map 13).

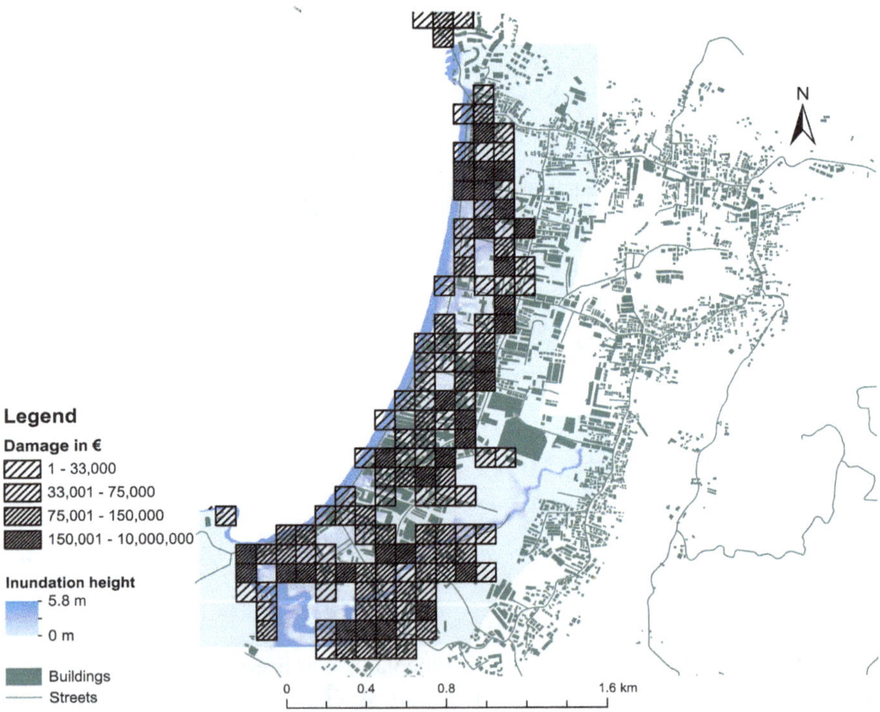

Map 13: Inundation height and accumulated damage in Patong Beach (Based on own calculation on TRAIT household and company survey and TRAIT tsunami simulation data)

No more than 15 % of all households live in the area that was inundated by the tsunami in 2004, and the majority of these households live in multi-story buildings, mostly on the upper levels, so that the highest share of physical damage resulted from minor damage to buildings (52 %) and to vehicles (36 %). Damage to household belongings or interiors only played a minor role, since it accounted for not more than 4 % of the total physical damage. On the other side, the damage occurring to businesses was also limited. In total, 30 million € in damages was counted for company property, while households suffered a total damage of roughly 20 million €. Of all communities, Patong Beach suffered from indirect effects for the shortest time. On average, the households reported five months of reduced income or unemployment after the event, which is significantly shorter than in the surveyed communities in the Phang-Nga Province. In the main sectors in Patong Beach - services and hotel and restaurant - people with contracts had a significantly longer reduction in income than people without contracts or self-employed entrepreneurs, which might be caused by the fact that people with contracts kept their original job, whereas the others could react more flexibly to the situation and find new fields of employment. Additionally, only 30 % of the population of Patong Beach experienced a loss of income. Nevertheless, this accumulates to a total amount of 10 million € for the workforce over the whole period of time. Despite the fact that physical damage to companies was also limited, the businesses in the area, which are highly dependent on tourism and the connected inflow of external spending power, reported long periods of reduced sales. Overall, the businesses in Patong Beach reported the shortest time for reaching the production level enjoyed before the tsunami. As previously stated, the businesses needed around one year to recover to normal in terms of their sales levels. Despite this, the deviation between the business sectors reveals that the hotel sector in particular needed significantly longer - more than two years - to reach its original production level. Other sectors that are more aimed at the local population or day-trippers, who came back to Patong first, needed less than one year to return to normal. Due to the high sales volume of the hotel sector - more than 4 million € on average for 2003 - the indirect losses for all businesses in Patong Beach add up to more than 350 million €, which is equivalent to 44 % of the total sales in the year after the disaster (cp. Table 56).

Table 56: Economic effects on household and companies (Own extrapolation)

	Households		Companies	
	Direct impact in million €	**Indirect effects caused by loss of income in million €**	**Direct impact in million €**	**Indirect effects caused by reduction of sales in million €**
Ban Nam Khem	6.2	6.5	5.3	6.1
Thai Muang	0.0	0.2	0.0	0.0
Khao Lak	6.6	19.5	90.3	5.7
Patong Beach	20.1	10.7	30.4	350.2

Overall, the total damage in Patong Beach adds up to around 45 % of the total damage that occurred on Phuket Island, while the indirect damage reached 42 % of the island's total indirect effects. This underlines the importance of Patong Beach for the island's economy and the local focus of the tsunami damage on several small spots in the province of Phuket. In the Phang-Nga Province, the three study regions represent nearly 60 % of the total damage incurred along the Andaman coastline in this province (NIDHIPRABHA 2007:20). This is particularly due to the fact that a large number of economic assets are concentrated in Khao Lak, and that this area was widely affected. In terms of indirect effects, the region accounts for 20 % of the total indirect losses, which is significantly lower than in Phuket, but underlines the lower economic importance of this region in comparison to Phuket, which is much further developed.

7.4 Conclusion

The previous discussion of influential factors and characteristics reveals several significant differences in the setting of the economic systems in the study regions surveyed. Without a doubt, the physical vulnerability of households and companies can be modelled using the combination of their exposed assets and their specific location on the micro-level (cp. Chapter 3.2.2). Against the background of the varying economic exposure and damage rates in the study regions, it is crucial to identify which characteristics determine the sensitivity of the economic system and which can be counted as the "system's ability to bounce back to a reference state after a disturbance" (TURNER et al. 2003:8075).

As seen above, the study regions chosen offer several significant disparities in their economic circumstances for studying the influence and interrelations of the economies' characteristics in the vulnerability framework. Besides the physical sensitivity or direct sensitivity of buildings, households or productive assets, monetary flows are also subject to the influence of the external shock. These elements of the economic system are especially determined by the economic linkages and the sources and destinations of the monetary flows. The economically linked study regions, with their focus on agriculture and tourism in the Phang-Nga Province and the almost exclusive focus on tourism and related businesses in the Phuket Province, provide the opportunity to study these impacts. In terms of this covariant and rapidly occurring hazard, these elements solely determine the influence of the impact on livelihoods and business activities. Because the sensitivity of exposed assets is determined by the location of monetary goods, factors deriving from the entitlement or the sustainable livelihood approach are categorized as resilience aspects allowing a faster recovery of the household or company. Here, the companies and households in the chosen study region offer a good insight into the varying circumstances of economic resilience elements. Due to the different development stages between the Phuket and Phang-Nga Provinces, households in these two regions differ in terms of their internal resilience capabilities. Variation among the communities can be found in differing income levels, income diversification and security of employment in terms of possessing a working contract or being self-employed. While in one community, an extension of indirect effects on the employment from informal employment could be found, in Patong Beach, persons employed without contracts could return to normal faster. No significant differences among the communities were found in terms of the access to the banking system or the education level. Due to the high variation coefficient among the households, significant differences were expected in the micro-level assessment. Significant differences on the community level in terms of resilience capabilities resulting from the entitlements of companies can be found in the sales level and employment level of the companies. It is assumed that larger companies have easier access to outside capital in order to cope with the physical impact, which can also be observed in the significant differences among the amount of credit and savings of the companies in the study regions. The relatively young economic development in Khao Lak and the dynamic business situation in Patong Beach result in a significantly younger population of businesses, which are assumed to be less experienced in their business activities, but also have a potentially more agile way of adapting to the shock. Since the companies in the region are strongly on tourism and agriculture and fishing in all of the communities, a dependency on seasonal changes throughout the year can be found. This might hinder the companies' coping abilities, especially when the business activities in the high season are blocked. In the following section, the specific characteristics of the businesses and households deriving from internal capabilities and external influences will be tested in their interrelation, determining the vulnerability situation in 2004 and the coping period which followed.

8 Interrelations within the vulnerability models

8.1 Households

The theory-based model developed in chapter 3 and the first descriptive analyses in chapter 6 and 7 lead to a combined vulnerability model linking the elements exposure, sensitivity and resilience (cp. Figure 30). This model should allow an evaluation of the processes and interdependencies determining the vulnerable situation of the households and their capabilities to cope with the shock.

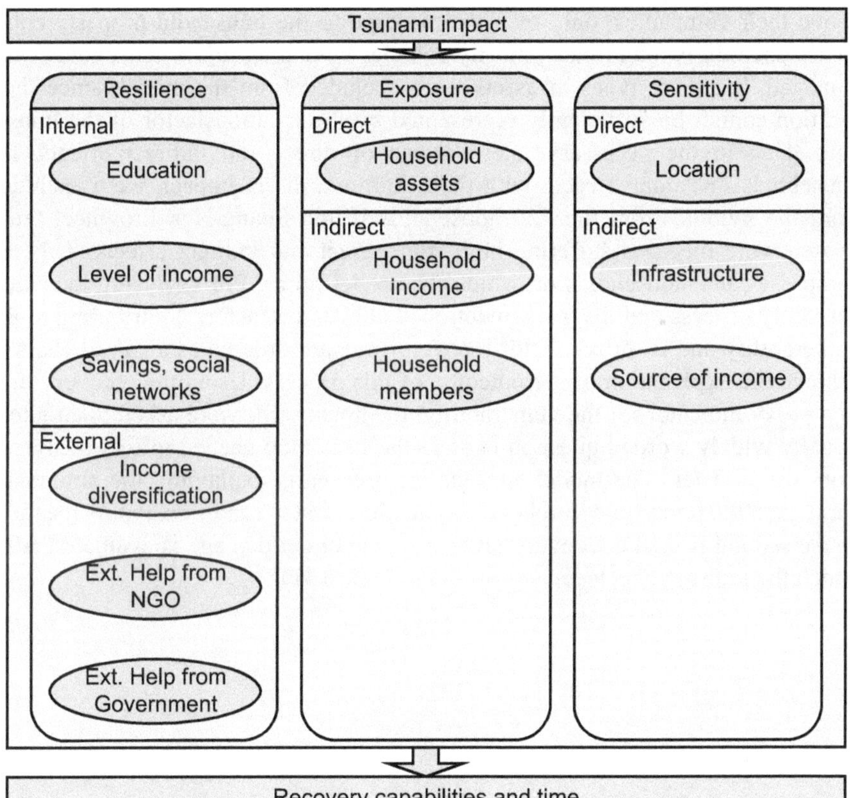

Figure 30: Framework for household vulnerability assessment (Author's design)

From the discussion in chapter two and three, the following hypotheses can be derived to test the relations between the single elements:

H_1 The level of direct damage caused by a tsunami increases vulnerability
H_2 The level of indirect effects caused by a tsunami increases vulnerability
H_3 Internal resilience capacities reduce vulnerability
H_4 External resilience capacities reduce vulnerability

From the 323 households interviewed in the four study communities, only the households in the province of Phang-Nga could be integrated into the further modelling due to restrictions in the chosen model. Since the majority of households in Patong Beach are located further inland or live in multi-story buildings above their companies, only marginal damage to the household property could be assessed. In Patong Beach, mostly vehicles belonging to the households were damaged, but these types of assets were excluded from the model, since their location cannot be sufficiently represented by the location factor of the household. Due to the necessary interrelations of direct and indirect effects, the households in Patong Beach, with their marginal direct impact, were excluded from the model. From the 224 households in the Phang-Nga Province, those chosen were the ones suffering both from direct and indirect effects. This encompasses 171 households on which the model relies. For analysing their vulnerability process and the goals mentioned above, the factors contributing to the vulnerability and its proxy factor recovery time are ordered as a causal chain to achieve the explanation of the outcome of this process. Using the recovery time as a proxy indicator for the vulnerability, the households were asked in an intentionally widely worded question to state the time they needed to feel recovered from the disaster. The model consists of three parts explaining the amount of damage in the household in relation to the pre-disaster condition and its location. In the second part, the recovery process of the direct damage is evaluated, after which the indirect effects are assessed (cp. Figure 31).

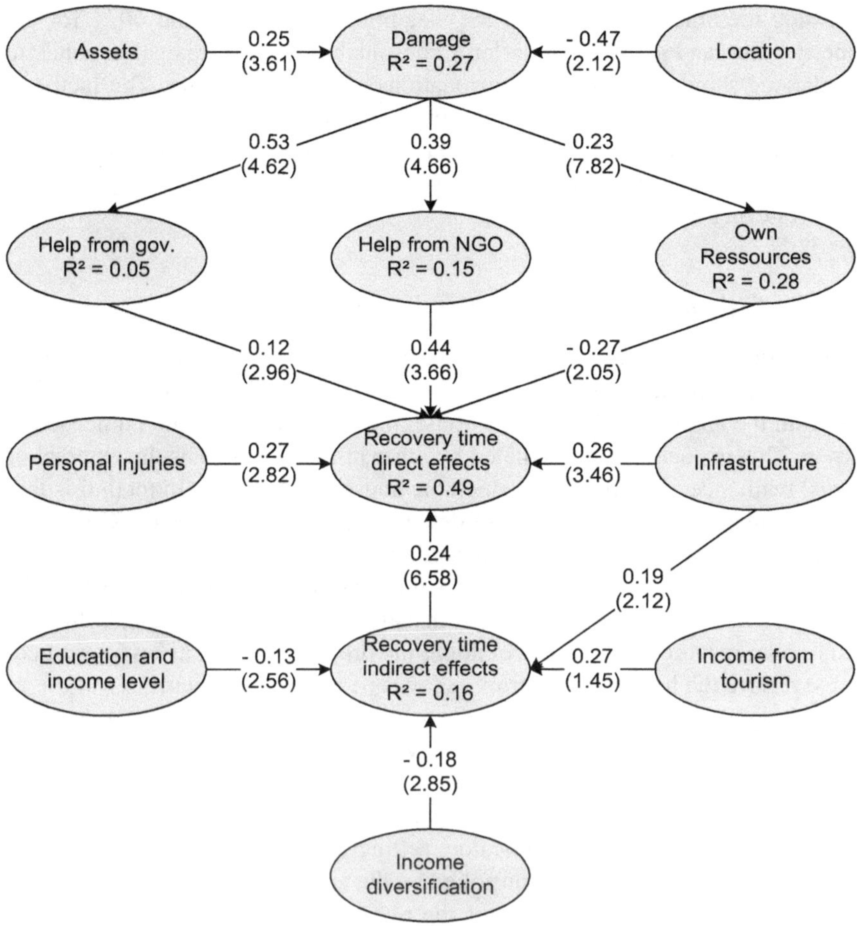

Figure 31: Structural equation model of household vulnerability (Own calculations based on TRAIT household survey – n = 171)

For the construction of the factors, the approach of formative latent variables was used. The constructs in this model represent explanatory combinations of indicators determined by their combination of variables. These factors are considered as index values representing their related indicators in the form of a linear combination (cp. Appendix - Table 70; DIAMANTOPOULOS and WINKLHOFER 2001:270). For example, the factor describing the amount of household assets is set up weighting the property value with three quarters and the long-term consumable goods with one quarter (cp. Appendix - Table 70). Regarding the factor describing the amount of damage, the weighting shifts to around 60 % for the property value and 40 % for long-term consumable assets. This corresponds to the relations shown in descriptive evaluations of the survey data. The factor of recovery time shows a contrary weighting indicating a higher value for the recovery time needed for long-term consumable goods than for property damage. This is explained by the fact that the households started to rebuild the houses before restoring their assets.

8.1.1 Relations in vulnerability elements

Focusing first on the factor describing the recovery time from the direct impact of the tsunami on the households, it stands out that there is no significant influence from the amount of damage the household suffered on the time it needed to recover. This impact is fully mediated by other influences such as the external or internal resilience capabilities on which the household can rely. In total, this factor shows an acceptable coefficient of determination with a corrected R^2 value of 0.49. A reducing effect on the time needed to recover can be found in the household's capability to use its own resources for repairing damage. Using the partial correlation method with the amount of damage as a control variable, both indicators - own resources used for rebuilding the property as well as own resources used to restore the household interior - show a negative coefficient.

Looking at the external resilience capabilities represented by the amount of help received by governmental or non-governmental agencies, it can be found that there is a positive influence on the recovery time, pointing out that the dependence on external help leads to an increased recovery time. Evaluating the relations between each of the indicators setting up the factors reveals differing effects. The partial correlation, controlled by the amount of damage, between the amount of external help provided for the recovery of personal belongings and household interior, and the time needed to recover these items, is negative. This points to a faster recovery of households that received external help for restoring personal belongings. On the other hand, looking at the time needed to recover from damage to buildings, the correlation shows a positive influence. Combining these two elements of household property affected by the tsunami into a single factor leads to an overall extending effect of external help on the recovery time following the greater weight that the damage to property value has been given by the algorithm (cp. Appendix - Table 70). Here, a strong f^2 value indicates a high influence of these two factors. Furthermore, the recovery time of direct effects is influenced by the effects of the damaged infrastructure, indicating that people from highly damaged villages were moved to temporary shelters while the infrastructure was reconstructed, extending the time period before the household can start to rebuild its own property. Personal injuries among the household members are also likely to extend the time period needed to recover from direct effects (cp. Table 57).

Table 57: Recovery time from direct damage (Own calculations based on TRAIT household survey)

	Path-coefficient	Indirect effects**	f^2	VIF	corr. R^2
Governmental support*	0.12	0.00	0.02	1.30	0.49
NGO support*	0.44	0.00	0.19	1.80	
Usage of own resources*	-0.27	0.00	0.09	2.50	
Disruption of infrastructure*	0.26	0.32	0.12	1.20	
Personal injuries*	0.27	0.00	0.09	1.10	
Indirect effects*	0.24	0.00	0.07	1.50	
Income diversification*		-0.04	0.00	n/a	
Damage		0.05	0.00	n/a	
Location		-0.02	0.00	n/a	
Income from tourism		0.06	0.00	n/a	
Education and income level*		-0.03	0.00	n/a	
Assets		0.01	0.00	n/a	

*significant min 5 % error- TRAIT Household Survey – Durbin-Watson: 1.72 – n = 171
Indirect effects are calculated as the difference of total effects and path-coefficient (direct effect)

According to SMITH and WARD (1998), effects of natural disasters on tangible assets can be divided into direct and indirect effects. Examples of indirect tangible effects of natural disasters include the disruption of supply chains or reduced income for households, even for households that are not directly affected by the natural disaster. One indirect effect that showed a significant influence on the recovery time of households is the loss of income after the tsunami. The duration of the influence of indirect effects on the household is mainly determined by household characteristics concerning its employment situation. Because the tourism sector suffered from a much higher loss and needed much more time to reach its original production level than other sectors characteristic for this area, such as agriculture and fishing, a significant impact can be found in the influence from this sector. A reducing effect on the indirect effects results from the diversification of income within a household. To measure this, the

Shannon Equitability Index was used, putting the number of income sources and their share of the total household income in relation and combining this in one index value (SCHWARZE and ZELLER 2005:56). The reducing effect of a diversified household income is higher than the result of a high level of total household income and education level. This factor is mostly influenced by the level of education among the working household members, while the total amount of income plays only a minor role (cp. Table 58). As seen in the previous evaluations, an unclear pattern evolves concerning the influence of the status of the employment of the household members. While in one community, household members with an official contract show an extended time with loss of income, in other communities, this was not the case. As a result, the model showed an insignificant influence of the type of employment, whether formal, informal or self-employed in one's own business.

Table 58: Recovery time from indirect effects (Own calculations based on TRAIT household survey)

	Path-coefficient	**f^2**	**VIF**	**corr. R^2**
Income from tourism*	0.27	0.08	1.04	0.16
Education and income level*	-0.13	0.02	1.06	
Infrastructure*	0.19	0.05	1.00	
Income diversification*	-0.18	0.03	1.09	

*significant min 5 % error- TRAIT Household Survey – Durbin-Watson 1.81 – n = 171

The duration of the recovery time from indirect effects is also determined by the disruption of infrastructure and its negative effect on the household livelihood. The duration of time needed to rebuild the road network and the water and electricity systems particularly affects the household's income situation. Since the majority of people possess a mobile phone, the telephone network could be fixed within two weeks. No extending influence could be found resulting from this indicator.

8.1.2 Discussion

The structural equation model developed for the households shows the relations and interactions between the three different components of vulnerability mentioned in the vulnerability frameworks discussed. Integrating all elements of vulnerability, it allows the quantifying of relations in the vulnerability framework and for an understanding and reconstruction of the status of vulnerability and the process of recovery that represents it in a most comprehensive way. The first key element of vulnerability, the exposure of household assets, could be strongly linked to the location of the household. The coastal terrain in the study areas of Khao Lak and Ban Nam Khem, with their flat areas stretching along the coastline and extending almost 2 km inland, contributed greatly to the total damage incurred by the households. On the other hand, the community of Thai Muang, protected from the direct wave impact by a peninsula that reduced the total damage per household, showed a similar recovery time to the highly exposed areas of Khao Lak and Ban Nam Khem, indicating a higher vulnerability level of this community. The majority of households are located in areas roughly eight metres above sea level, which is still a tsunami-endangered zone according to the 2004 tsunami wave heights ranging from 5 metres in Thai Muang to 12 metres in Ban Nam Khem and Khao Lak (DDPM 2007; cp. Chapter 3.2.2). Households which had to rebuild their property suffered from a loss of personal belongings of up to 8,000 €, while households whose houses were able to withstand the wave impact reported a loss of about half this figure.

The road system and lifelines such as the water system showed a particularly extending influence on the household's recovery time from both direct and indirect effects. The road system in the Phang-Nga province was severely damaged by the tsunami, needing an investment of more than 2.6 million € (NIDHIPRABHA 2007:5). Due to the enormous efforts undertaken by the government and the Thai military, the road system was restored after two months and the telecommunication network after only two weeks (TRAIT Household Survey). In some parts of the study region, it took up to 2.5 years before all infrastructure was restored and household members could reach their workplaces and manage their everyday life without limitations.

As introduced by TURNER et al. (2003) in their vulnerability framework and picked up by CALGARO and LLOYD (2008) examining the vulnerability of Khao Lak, the sensitivity of the household livelihoods in this region is notably determined by the limited livelihood options. The focus on tourism in the core region of Khao Lak and its surrounding villages in particular and the resulting lack of income diversification leads to greater sensitivity. Concerning these indicators of vulnerability, it was found that there is no direct linkage to the recovery time of household assets. These indicators are mediated by the indirect results of the tsunami, especially the loss of income, but still show a significant effect on the outcome of vulnerability. Findings from the survey show that the households in the region generate between around 28 % (in Thai Muang) and 44 % (in Khao Lak) of their income from tourism or related businesses. As a result, households which were not dependent on tourism recovered on average one month earlier. Households which generate a higher share of income from tourism-related industries show a longer time with losses of income than other households. From the theory of vulnerability, the hypothesis was derived that internal as well as external resilience capacities reduce the vulnerability of households, and therefore decrease the time needed to overcome the effects of this natural hazard. The structural equation model indicates that at least the external resilience capabilities, represented by the help provided by non-governmental and governmental organisations, lead to an increased recovery time. The government paid compensation of nearly 45 € to all individuals affected by the tsunami, regardless of the amount of damage or any other circumstances of these individuals. If the house had been destroyed completely, a compensation of 430 € was paid. NGOs did not follow this restriction and provided an amount of compensation per household which was nearly three times higher. The tsunami-related funding in 2004 broke many records and reached nearly 10 billion € from international sources. As a consequence, the recovery process in the tsunami-affected countries could be financed without any budget constraints (FLINT and GOYDER 2006). Due to findings of the 'Tsunami Evaluation Coalition', the activities which took place during the recovery process showed a lack of assessment, and the international response was 'insufficiently evidence-based' (DE VILLE GOYET and MORRINIÈRE 2006:13). The report focuses on the poor coordination and the lack of exchange of assessment reports between the national coordination and international agencies. This lack of coordination resulted from the large number of players that were involved in the recovery process. Furthermore, the national government of Thailand declined international help. As a consequence, only international agencies that had had partnerships with local organisations before the disaster could act in southern Thailand. Therefore, a huge number of organisations were involved in the recovery process which were not specialised in this kind of event, such as religious or missionary organisations (KORFF 2009; cp. Chapter 6.4). Governmental coordination that would have been necessary to direct the recovery efforts lacked knowledge, skills and adaptive capacities (KAMOLVEJ 2006). The recovery time was thus extended. Due to the competition among the donors, an oversupply of help among the population could also be seen (TELFORD and COSGRAVE 2007:4). This reduced self-initiative led to a dependency on external help (KORFF 2009). The structural equation model showed that households which were able to finance and organise their own recovery process were able to return to normal significantly faster.

8.1.3 Conclusion

The structural equation model developed shows the relations and interactions between the three different components of vulnerability. Using data from the study area in Khao Lak, it allows us to quantify these relations in the vulnerability framework. The model showed that the first hypothesis stating that the amount of direct damage caused by the tsunami has a significant influence on the recovery time needs to be rejected. A special meaning is taken on by the external help provided by government and NGOs, which showed an extending effect on the recovery time. Therefore, the hypothesis concerning this element of resilience needs to be rejected as well. This may be caused by the special situation in southern Thailand after the tsunami, which is less comparable to other natural hazards, resulting from the large amount of donations and the unprepared coordination and allocation of supporting measures. Internal resilience capabilities positively influence the recovery from direct and indirect effects. Here, the indirect effects mediate some of the internal resilience capabilities, which show a significant and strong impact on the households. Further research should focus on the question of whether the lessons learned from this major natural disaster match the ongoing adaptation strategies in the region, and especially whether the problem is being addressed of the mono-structured and less diversified region, which increases the vulnerability of households, in order to reduce the population's vulnerability and enable a faster recovery process.

8.2 Companies

From the total of 295 companies interviewed, not all could be included in the further investigation. The only companies included in the modelling of the recovery and vulnerability processes were those that experienced direct impacts of the tsunami, i.e. physical damage to company property, and indirect effects, such as the reduction or loss of sales, or difficulties in production caused by problems with suppliers, infrastructure or workforce. 199 companies, or 67 % of the total number interviewed, experienced tsunami impacts on their business activities. These 199 companies will be the further basis for all evaluations. The distribution between the four study communities is uneven in terms of the percentage of affected companies. While in Ban Nam Khem, all companies interviewed were directly affected by the Tsunami, the number differs significantly for the other communities. In Thai Muang, only one-third of all companies experienced direct effects, while in Khao Lak and Patong Beach, the figure was 60 %. Companies not affected by the tsunami impacts were either located outside the inundation zone or not linked to affected businesses, or were established after the disaster. This leads to a methodological problem of a post-disaster field survey where it is not possible to interview companies that are out of business due to the consequences of the disaster. This is important to highlight because as a consequence of this fact, the most vulnerable businesses, i.e. those which were forced to close or to relocate their branches, are excluded from this study. Nevertheless, the model provides an insight into the processes that took place in the period since the impact and attempts to represent the process of recovery. In the hypothetical case that these companies could have been included, the model might show even stronger relations within the single influence factors, especially for the companies that had to stop their business activities because of the influence of consequences of the tsunami or shortages in the access to assets in the manner of the sustainable livelihood approach.

8.2.1 Relations in vulnerability elements

The modelling approach used in this study follows the three elements of vulnerability defined by TURNER et al. (2003) or the BBC-Framework (DE VILLAGRÁN LEÓN 2006), namely the exposure, the sensitivity and the resilience part of a vulnerable system. Based on the literature review and the conceptual framework for this study in chapter three several indicators and their hypothetical influence could be identified representing the single parts of vulnerability (cp. Table 4; Figure 32). Due to the aim of the chosen structural equation model approach using the partial least square algorithm as a prognostic tool, which attempts to create a model best representing the interrelations within the given dataset, indicators that do not contribute significantly to the model need to be deleted from the model in order to create a valid measurement (WEIBER and MÜHLHAUS 2010:209).

For the representation of the different perspectives of the recovery process, formative constructs were chosen instead of reflective constructs created by factor analysis. Based on the theoretical discussion in Chapter two and three, the following model could be derived. Additionally, theoretical assumptions can be set up based on this model approach:

H_1 The level of direct damage caused by a tsunami increases the vulnerability
H_2 The level of indirect effects caused by a tsunami increases the vulnerability
H_3 Internal resilience capacities reduce the vulnerability
H_4 External resilience capacities reduce the vulnerability

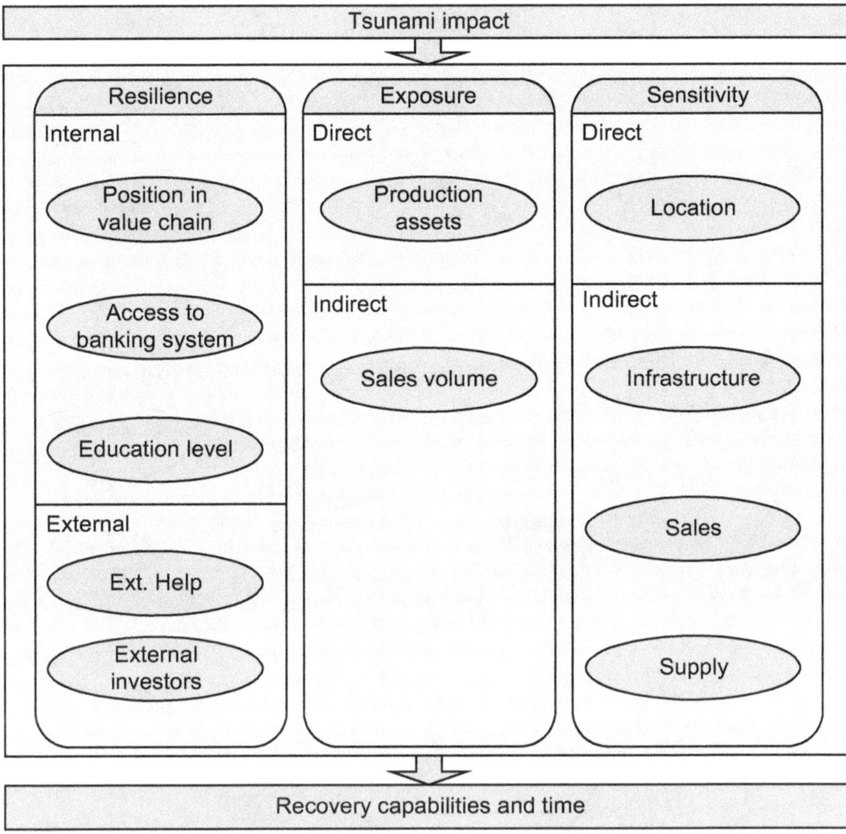

Figure 32: Framework for company vulnerability assessment (Author's own design)

To achieve a vulnerability model for the companies parallel to the household situation, four stages in the causal chain of explaining the process and determinants of the recovery period can be distinguished. The recovery time required for a complete - and for the company sufficient - recovery was used as a proxy indicator for the company's vulnerability. In the first step, the model explains the amount of physical damage in the company based on its pre-disaster condition and its location. The recovery process of the physical damage is then evaluated. In the following section, the recovery from indirect effects regarding the connection to suppliers and customers is linked to the overall vulnerability index.

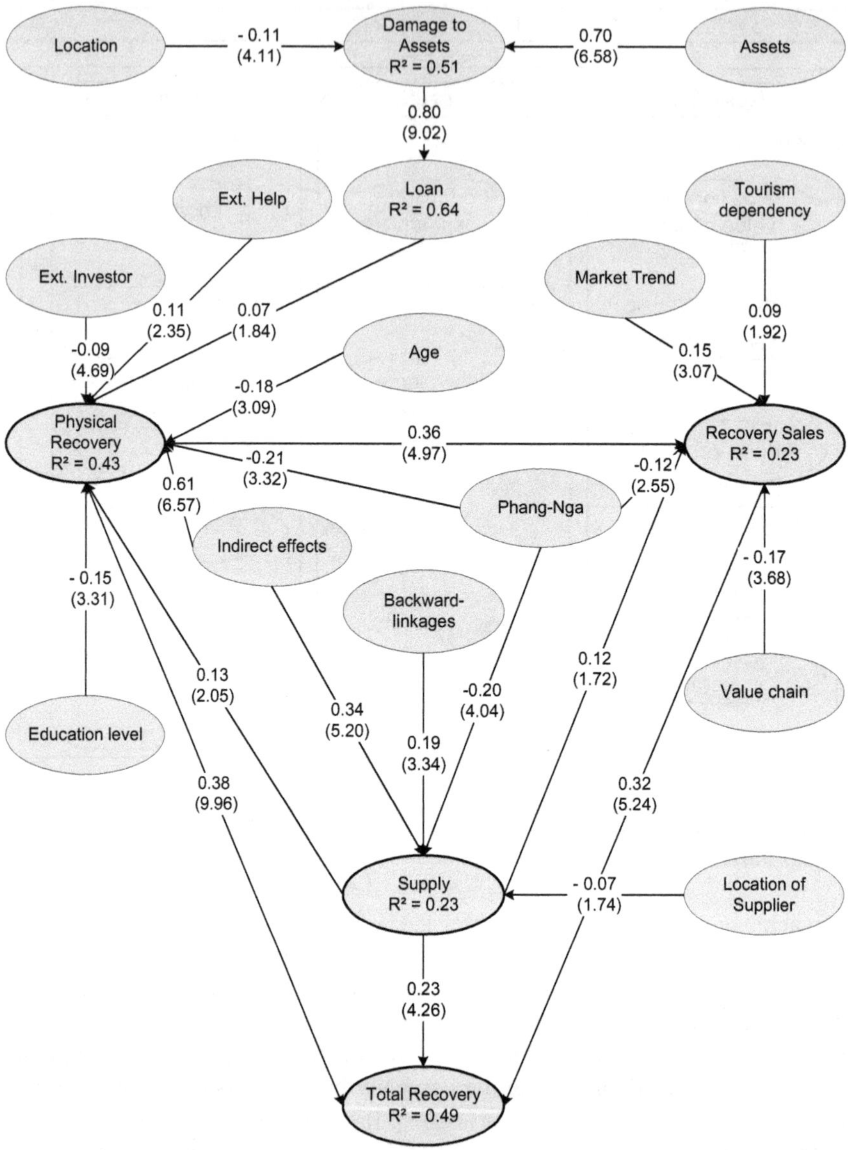

Figure 33: Structural equation model of the company vulnerability (Own calculations based on TRAIT company survey – n = 199)

8.2.2 Discussion

The investigation into which factors influence the duration of the recovery period was driven by the outline set through the vulnerability framework from TURNER et al (2003) defining the status of vulnerability as influenced by the three perspectives of exposure, sensitivity and resilience. Besides the notion that each perspective has a direct influence on vulnerability, it points out there are also strong interrelations between the three components. In accordance with this THYWISSEN et al. (2006) defined their equation of vulnerability, which expresses the statement that exposure and sensitivity can be reduced or completely compensated for by the resilience perspective that is available in the system. To reproduce and model these circumstances of the vulnerability process, the data were tested for direct and indirect linkages of the factor 'total damage to company assets' with the derived proxy of physical vulnerability being the 'recovery time from physical damage' to company assets. Comparable to the situation of the tsunami-stricken households in the Phang-Nga and Phuket Provinces, the direct correlation between these two items shows only a weak significant correlation on a 10 % error level with a positive extending influence when applying an OLS-Regression or a bivariate correlation. As a consequence, this relation must be mediated by other factors interfering in the statistical interrelation. Due to the fact that governmental and non-governmental agencies attempted to speed up the recovery process by donating assets or giving credits at reduced rates of interest to the affected companies, these items must be taken into consideration when modelling the recovery process (cp. Chapter 6.4). Additionally, the capacities within the company, previously defined as internal resilience capabilities, need to be tested for their influence on the recovery from the physical impacts. External support in the aftermath of the disaster was split into two parts. Physical goods were donated to micro-enterprises mainly from the business sector of agriculture and fishing, while the medium and large companies, mostly from the service and hotel and restaurant sector, were supported with low-interest credits (cp. Chapter 6.4.1). These two factors for the external financial resilience capabilities were tested for statistically significant relations mediating the recovery time in relation to the amount of physical damage for the companies. It showed that only the relation between the amount of physical damage and the amount of credit granted to the company reveals a significant correlation on the level of 0.8 with a 1 % two-sided error level. The external help received stands in no significant relation to the amount of physical damage at all. The reason for this is that the focus of the donation of physical goods was placed on micro-enterprises from agriculture and fishing which were, in some cases, over- or under-compensated for (TRAIT Company Survey 2009; FG-Fish-KL; FG-Fish-BNK).

One's own resources in the form of savings, insurance claims or substitutions from social networks (e.g. family, friends etc.) also showed no significant influence on the duration of the recovery period and no significant correlations to the total amount of damage, since some companies, especially in Patong, where only minor damage occurred, used the occasion to invest more money for business improvement purposes.

As a result of this situation, the total amount of damage is regressed in the model on the credit volume of the company for the year 2007. The corrected R^2 of 0.64 indicates a close fit of the regressing variable. Besides the direct regressor, the upstream factors, which are the pre-tsunami assets and the location of the company, also show indirect effects on the amount of credit the company had after the disaster. Similar to the influence on the variable 'tsunami damage', the pre-tsunami assets also have an increasing influence, while on the other hand, the location of the company has a decreasing influence. Since the location factor for the company is set up by the indicators 'height' and 'distance', higher values indicate a safer location, while smaller values point to a more physically exposed location of the company (cp. Table 59; cp. Appendix - Table 71).

Table 59: Explanatory model for the amount of loan (Own calculations based on TRAIT company survey)

	Path-coefficient	**Indirect effect****	**f^2**	**VIF**	**corr. R^2**
Direct effects*	0.80	0.00	1.76	n/a	0.64
Assets*		0.56	0.00		
Location*		-0.09	0.00		

* significant min 5 % error – TRAIT Company Survey – Durbin-Watson: n/a
** Indirect effects are calculated as difference of total effects and direct effects (path-coefficient) – n = 199

In the next step, the model aims to explain and reproduce the recovery time for the physical impact of the tsunami disaster. As external financial influence factors, both the amount of credit and the external help in terms of donated goods were taken into account. Both of these external measures for coping with the consequences of the natural disaster show positive path coefficients, which points to an extension of the recovery time for companies receiving higher external financial or physical support. This fact is parallel to the situation found for the households, where external help leads to an increase in recovery time due to the dependency in granting the financial support, with 'aid dependency' sometimes following (cp. Chapter 6.4.1). The external help was provided in the form of the direct donation of goods, i.e. occupational tools, boats and vehicles, or the provision of earmarked funds for the purchase of the specific goods by governmental or non-governmental agencies. The path coefficient of 0.11 with an f^2 value of 0.02 shows an acceptable but low contribution to the factor. These results correspond to the extending influence on the recovery time which could be observed for the households in the specific structural equation model in terms of explanatory power and path coefficient. Explanations for this extending effect can be found in the discussions and outcome of the focus groups. In nearly all of the focus groups for the different occupational groups of the study area, complaints were made about the uneven distribution of relief goods which "lacked systematic management" (FG-Fish-KL). The focus groups also reported cases of over-compensation, especially in the fishery sector, where higher pre-tsunami assets and losses were reported to the authorities than were actually possessed/sustained by the entrepreneurs. In other cases, the business owners were not able to apply for the support due to traumata (FG-Inc-BNK), or were overstrained by the bureaucratic procedure (FG-Tou-PB).

While these measures were mostly aimed at small-scale enterprises, the government tried to boost the general economic situation in the region by supporting medium and large companies in particular with low-interest credits (Table 60).

Table 60: Loan and external help to companies (Own calculations based on TRAIT company survey)

	Loan in euro		**External help in euro**	
	Mean	**Variation coefficient**	**Mean**	**Variation coefficient**
Micro enterprise	$6,111_a$	2.35	568_a	0.85
Small enterprise	$17,500_a$	2.02	$5,400_a$	1.34
Medium enterprise	$1,111,939_b$	1.85	0_b	n/a
Large enterprise	$131,123_b$	2.00	0_b	n/a

Subscripts in the same row not sharing the same letters are significantly different at $p<0.05$ in the two sided t-test - $n = 199$

The physical recovery for the companies comprises the period of time needed to rebuild or restore the production assets of the company. In this model, these assets are represented by property value, machinery and other equipment. Following the conceptual framework of this study, the physical recovery depends on the availability of internal and external assets on which the company can rely within this process.

First of all, the companies' recovery process is strongly limited by the degree of devastation of infrastructure in the surrounding area. This influence can be quantified with a large f^2 effect size of 0.63 dominating the whole recovery process. This indirect influence comprises the degree to which lifelines such as telephone, roads, electricity and water were affected, as well as limitations arising from problems in terms of labour supply when employees were killed or injured by the wave impact, or had just been unable to come to work due to infrastructural problems. These indicators jointly formed the factor of 'infrastructural recovery', since the expressions of these variables are nearly parallel. Values differ between one week in Thai Muang, one and a half months in Patong Beach, and around three months in Khao Lak and Ban Nam Khem. A common feature of both of these indirect effects is that the company itself only has limited influence on the progress of solving this problem. None of the companies reported initiating measures to fix the problems with the infrastructure hindering their business, and less than 6 % of the companies reported starting their own initiatives to find new employees directly after the wave impact. Another common issue of these indirect infrastructural effects is that they are a crucial basis for further recovery activities, which expresses itself in the fact that these two recovery indicators show the shortest time of all, which indicates that the basic infrastructural problems need to be solved before the reconstruction of buildings and the business activities themselves can be started. Furthermore, the indirect effects generate the highest explanatory power for the 'physical recovery' of the companies (cp. Table 61). The f^2 value reaches 0.63 and indicates a strong and outstanding explanatory effect. The duration of the infrastructure recovery (road system, electricity, water, telephone) greatly extends the physical recovery time with a path-coefficient of 0.63. Since the level of damage shows no direct influence, but is mediated by the amount of credit received by the company, it only has an indirect effect on the physical recovery period, which is not traceable in terms of the effect size. The location of the affected company also has no measurable influence on the duration of physical recovery. This points to the fact that factors deriving from the sensitivity or resilience dimension of vulnerability are more important in terms of vulnerability than the amount of the physical impact.

In terms of external resilience and recovery capabilities, the external help provided in the form of benefits in kind, such as productive assets concerning vehicles, fishing gear or other occupational tools, leads to the highest influence, in terms of the f^2 value, on the recovery time. Parallel to the situation in the household vulnerability, the dependency on receiving in-kind benefits also leads to an extension of the physical recovery time. Similar effects can be seen in the outcome of the low-interest loans for the company recovery. In this case, the effect size is slightly smaller than for the in-kind benefits (cp. Table 61).

Table 61: Physical recovery (Own calculations based on TRAIT company survey)

	Path-coefficient	Indirect effect	f^2	VIF	corr. R^2
Loan*	0.07	0.00	0.01	1.09	0.43
Education level*	-0.15	0.00	0.03	1.09	
Ext. help*	0.11	0.00	0.02	1.03	
Indirect effects*	0.61	0.00	0.63	1.05	
Age*	-0.18	0.00	0.04	1.03	
Ext. investor*	-0.09	0.00	0.01	1.03	
Phang-Nga*	-0.21	0.00	0.03	1.03	
Assets*		0.04	0.00	n/a	
Direct effects*		0.05	0.00	n/a	
Location*		-0.01	0.00	n/a	

* significant min 5 % error – TRAIT Company Survey – Durbin-Watson: 1.73
** Indirect effects are calculated as difference of total effects and direct effects (path-coefficient) – n = 199

The overall level of damage strongly varies between the two study provinces. Since only a small strip along the coastline was affected in Patong Beach, the area could be cleaned from debris after only four weeks. Businesses located more than 150 metres from the beach line could reopen only several days after the event (TRAIT Company Survey; MYDANS 2005). After three weeks, 90 % of the impacted hotel rooms were available again and most hotels had reopened (TRAIT Company Survey; BIRKLAND et al. 2006). After six months, the recovery of the infrastructure had been completed (TRAIT Company Survey). Khao Lak, on the other hand, was heavily hit and almost completely destroyed along the coastline. As a consequence, the inflow of external funds from tourism shrank significantly and in 2005, the tourism revenue reached not more than one-fifth of the previous year (ILO 2005). The fact that it still reached 20 % is due to the large number of international volunteers that needed accommodation and food (TRAIT Company Survey). The rebuilding of the businesses was undertaken in stages depending on the availability of funds from NGOs or the government and the low-interest credits, and also the slowly increasing tourist inflows (CALGARO et al. 2009:23). This fact is also represented in the structural equation model by the introduction of a dummy variable for the location of the business either in the Phang-Nga or the Phuket Province. This variable indicates a significantly faster physical recovery for businesses in the Phuket Province, with a path-coefficient of 0.21 and an effect size of 0.03. This effect is clarified by separating the two provinces for further investigations using a correlation analysis between the amounts of damage incurred by the company and the time needed to restore the damaged building or production assets. For companies in the Phang-Nga Province, no significant correlation can be found, while for Phuket Island, the correlation coefficient is highly significant[5]. Additionally, and in contrast to the structure equation model for all companies, the amount of credit received by the company and the duration of rebuilding is not significantly correlated for companies in the Phang-Nga Province, while on the other hand, a highly significant correlation can be found for the companies in the Phuket Province[6].

The reason for the non-significant relation between the amount of damage and/or loan and the recovery time for the Phang-Nga Province must be seen in other hindering factors possibly deriving from the demand situation as described by CALGARO et al. (2009) or the supply situation or other vulnerability factors resulting from the resilience perspective. In Phuket, however, a significant correlation between the amount of damage and the amount of credit can be found, and it can be concluded that the governmental programme of providing low-interest loans to the affected companies was more effective here. In the focus group discussion organised by the TRAIT project, the majority of the participants from Patong Beach stated that they had benefited from the low-interest programme and reopened their businesses soon after the event, but were still facing a decrease in customer numbers (FG-Tou-PB). As discussed previously, the responsibility for implementing disaster management strategies is directed to the provincial and sub-district government level in order to guarantee that local conditions and needs are better catered for (SINGBUN et al. 2008). While the in-kind benefit aid was aimed at agricultural and fishing livelihoods for self-employed persons or micro-companies, the small and medium-sized tourism businesses did not benefit from this programme (TAN-MULLINS et al. 2007; FG-Hot-KL). Small, medium and large enterprises were supported by tax reductions or reduced interest credit programmes from the Tsunami Recovery Fund supported by the Venture Capital Fund of the Bank of Thailand. The programmes were particularly ineffective for smaller companies due to the bureaucratic procedure and, in some cases, also the fear of the business owners of taking out additional loans on their business while facing an uncertain future for the development of their sales activities (FG-Tou-PB). A reason for the differing effec-

5 For companies in the Phuket Province, a correlation coefficient of 0.46 with a p-value of 0.00 can be found while the affected businesses in the Phang-Nga Province show a correlation coefficient of 0.10 with a p-value of 0.46.

6 For companies in the Phuket Province, a correlation coefficient of 0.37 with a p-value of 0.004 can be found, while the companies in the Phang-Nga Province show a correlation coefficient of 0.07 with a p-value of 0.59.

tiveness in the aid programmes can also be found in the provincial responsibility for risk management. For Patong Beach, the recovery process was strongly supported by the Phuket Tourism Association, which, due to its economic importance, has strong connections with the Governor of Phuket and the national authorities. Besides the arrangement of additional funding for the island's rebuilding process, the association initiated the establishment of the Phuket Small Business Recovery Centre. The main task of this centre was to link the affected companies with government departments and other funding agencies to allocate funding opportunities, especially to smaller businesses for the rebuilding or restoration of damaged buildings, and also to support unemployed persons with their job search and social security applications (LARSEN et al. 2011:7). In Khao Lak, on the other hand, business owners mostly stated that "despite plans and promises, no concrete support measures such as a circulation fund have been established [by the government] to help small business entrepreneurs at all" (FG-Hot-KL). In Khao Lak, the Phang-Nga Tourism Association attempted to use its connections to the provincial government as well as in Patong Beach to complain about delays in funding and rebuilding. Additionally, the Khao Lak Small and Medium Enterprise Group was established to combine the political influence of the small businesses on the government in order to raise additional funding (LARSEN et al. 2011:6). As seen in the structural equation model and the comparison, this effort had not been as effective as in Patong Beach. Furthermore, it seems that in Phang-Nga, other influencing factors must have had a stronger impact on the physical recovery than the funding provided by the government and NGOs.

An aspect directly linked to the discussion above is the companies' internal coping capacities, which enable a faster physical recovery. Particularly in the context of the low-interest credits, the education level of the management plays an important role[7]. Comparing the two provinces of Phang-Nga and Phuket, it becomes obvious that in Phang-Nga, the correlation coefficient of 0.34 is significant (p-value of 0.01), while in the Phuket Province, no significant correlation can be found (correlation coefficient of 0.11 with a p-value of 0.40). Testing the influence of the size of the company, the variable 'pre-tsunami sales' was introduced in the comparison to exclude the effect of larger companies needing more financial capital. In the Phang-Nga Province, this reduced the correlation coefficient slightly to 0.31 (p-value of 0.03) and indicated that the education level of the management played a significant role in the application process. In the Phuket Province, on the other hand, the correlation coefficient remains insignificant,

7 The amount of credit and the education level of the company's manager show a correlation of 0.18 with a p-value of 0.06.

which leads to the assumption that the mediating agency in Phuket, the "Phuket Small Business Recovery Centre", could effectively support small businesses in the application procedure for low-interest credits and other relief programmes. In contrast, the in-kind benefit support from governmental or non-governmental agencies has no significant correlation with the education level at all, which therefore suggests that this kind of support was given to companies with fewer bureaucratic procedures and also to micro-enterprises or self-employed persons with a lower education level.

In addition to the interrelation with the external support, the educational level of the company itself influenced the duration of the physical recovery in a reducing manner. In the structural equation model, this is expressed by a negative path-coefficient and a low, but significant, effect size of 0.03 (cp. Table 61). Companies that can rely on the 'asset' of a higher education level can benefit from this by finding other financial sources for the recovery process in terms of ordinary bank credits or a generally more flexible adaptation to the new situation than others (WEBB et al. 2002). Additionally, the age of the company shows a significant and relatively strong influence on the physical recovery of the company. The negative path coefficient indicates that older companies are able to recover faster than younger companies. This stands in contrast to the study of WEBB et al. (2002), who found that younger companies were able to recover faster following the Northridge earthquake. They argued that older companies are less flexible in adapting to the new circumstances and are stuck in old behavioural patterns. The findings by WEBB et al. (2002) are valid for the whole recovery period including physical restoration and the rehabilitation of the sales level. In the model developed in this study, only a relation between the age of the company and the duration of the physical restoration could be verified. Results from the companies in the study area show that older companies in particular relied less on governmental or non-governmental programmes for financing and could go back to other funding sources such as social networks or other sources[8]. In contrast to this, there is no positive influence on the possibility of the company finding new customers based on experience in doing business. Another reducing influence can be found in the availability of an external investor, which, obviously, can easily initiate external inflow of funding for the recovery process. Such external investors can be found only in Patong Beach and Khao Lak. But even in these two communities, only around 10 % of the companies

8 Using the Kruskall-Wallis Test to compare the age of companies that financed the restoration of their company buildings with external financial support from the government or NGOs with the ones who reported bank credits or 'other resources' such as networks reveals that these companies are significantly older.

have an external domestic investor and another 5 % an international external investor. A closer look reveals that it is almost exclusively larger companies which have external investors: all of the large companies have either domestic or international investors, one-third of the medium-sized companies have a domestic and another 15 % an international investor, while only 15 % of the small companies have a domestic investor and only 2 % of these companies have an international investor. In the case of micro-enterprises, 3 % of the companies reported having external international investors, which results from the fact that these 'foreign investors' run a small-scale family business together with their Thai wife (TRAIT Company Survey). Businesses run by non-locals mostly did not benefit from the recovery programmes (CALGARO et al. 2009:113). Despite the fact that one-third of the affected companies had insurance coverage for parts of their damage, this element did not show an effect on the physical recovery time. Though the relation is significant, the path coefficient is too low to be considered, and the effect size is zero. This indicates that insured companies did not benefit from the risk transfer to an external company in terms of a faster recovery compared to the governmental support or non-governmental support programmes. Nevertheless, companies that covered their losses with insurance show a slightly lower credit volume and are therefore less vulnerable to future external shocks (TRAIT Company Survey).

In all four study communities, the companies felt a major constraint in their business activities over and above the direct physical impact on their property resulting from the reduction in sales. Business constraints resulting from the loss or interruption of supply ranked third, except in Ban Nam Khem, where the fishing companies and the industries depending on agriculture and fishing suffered from a shortage in supply and ranked the supply problems as the second most important issue (cp. Table 62).

Table 62: Percentage of affected companies ranking indirect effect 'very high' (Own calculations based on TRAIT company survey)

	Supply	Sales	Indirect effects
Ban Nam Khem	$19.5\%_a$	$63.5\%_a$	$17.0\ \%_a$
Thai Muang	$14.3\%_a$	$70.0\%_a$	$15.7\ \%_a$
Khao Lak	$12.5\%_a$	$68.3\%_a$	$19.2\ \%_a$
Patong Beach	$5.6\%_a$	$67.4\%_a$	$27.0\ \%_a$

Chi-square test indicates that distributions result from the same basic population and are similar - N = 199

The recovery of sales activities is determined by several factors which result from the particular circumstances of the target market of the company. In general, the business sector is referred to as the major reason for sales reductions after an external shock. Despite using the concrete business sectors of the companies, the modelling approach used in this study aims to integrate parameters representing the circumstances of the business sectors. Parameters describing the status of the company within the value creation chain, the dependency on cash flow from outside the region in terms of tourism dependency, the market seasonality of the targeted market and the region of the target market were all integrated into the model (cp. Table 63).

Table 63: Recovery of sales (Own calculations based on TRAIT company survey)

	Path-coefficient	Indirect effect**	f²	VIF	corr. R²
Phang-Nga*	-0.12	0.00	0.01	1.48	0.23
Tourist dependency*	0.09	0.00	0.01	1.25	
Value chain*	-0.17	0.00	0.03	1.17	
Market trend*	0.15	0.00	0.03	1.15	
Physical recovery*	0.36	0.00	0.10	1.33	
Indirect recovery*		0.22	0.03		
Education level*		-0.05	0.00		
Age*		-0.06	0.00		
Location*		0.00	0.00		
Backward-linkages*		-0.02	0.00		
Ext. help*		0.04	0.00		
Investor*		-0.03	0.00		
Loan*		0.02	0.01		
Assets*		0.01	0.00		
Direct effects*		0.02	0.00		

* significant min 5 % error – TRAIT Company Survey – Durbin-Watson: 2.20
** Indirect effects are calculated as difference of total effects and direct effects (path-coefficient) – n = 199

Despite its non-direct influence on the recovery of sales, the indirect effects in terms of interruption of lifelines and the unavailability of employees shows a strong effect on the recovery of sales. The duration of time for the clean-up after the event certainly played a significant role, during which the economic life in the region nearly stopped. This should also partially be found in the influence of supply on the demand recovery. In reality, this problem was solved after two weeks for the companies in the Phuket Province and after not more than three months in the Phang-Nga Province, where the duration of recovery time for companies was much longer. As a consequence, the interruption of supply shows a misleading positive path-coefficient indicating an extension of sales recovery in the case of longer supply interruptions. Due to the relatively limited tsunami-affected area and the focus of the companies on suppliers from the untouched regional centres, i.e. Phuket Town or Thai Muang (cp. Chapter 7.2.2), the supply problems are almost perfectly significantly correlated with the interruptions of infrastructure.

An important factor for describing the characteristics of the companies' sales activities can be extracted from input-output statistics and their figures, i.e. the backward and forward linkages which indicate a company's embeddedness in the economic value-adding circle (cp. Chapter 7.2.4). Both indicators show a positive loading, and therefore the factor represents the integration of the company into the economic value-adding circle in terms of both the backward and the forward linkages. As the path-coefficient of this factor, regressed on the duration of sales losses, is negative, this relation indicates that companies that are more strongly embedded in the economic cycle could cope better with the breakdown of customers than companies depending on final demand, such as retail, hotels or restaurants, and in restricted terms also the service sector. The agriculture and fishing sector and the manufacturing sector show particularly high values in terms of forward and backward linkages, which allows them to find new markets for their products more easily. In conclusion, it can be stated that these business sectors are less vulnerable and more capable of coping with the impact of external shocks, especially when they are affected only in indirect ways. Companies who show higher values in terms of their embeddedness in the value chain reported significantly more often using opportunities available by finding other markets and/or companies for their products in order to recover faster. However, the correlation between the sales volume in 2003 and the statement of finding new customers or markets on one's own initiative indicates that larger companies in terms of sales were particularly able to use this opportunity[9]. In addition to this, indirect effects can be found in the structural equation model resulting from the indicators 'education level' and 'external investor', which have a lowering effect on the duration of decreased sales. Using a correlation analysis shows that these company characteristics are able to increase the likeliness of proactive measures in the company aimed at finding new customer groups for their products.

Companies focused on the local market, especially on final demand from households, are naturally limited in terms of falling back on other regional markets and are especially prone to larger devastations. This fact can be seen by looking at the variable introducing the differences between the more extensively devastated areas of the Phang-Nga and Phuket Provinces. As already discussed in the previous chapter, the companies in the Phang-Nga Province needed significantly longer to recover to their original sales volume. As well as the dependency on external cash inflow, this effect is a result of the focus on the local popula-

9 The correlation coefficient between sales volume 2003 and a dummy variable representing companies' own initiatives for finding new customers (1 means a company's own initiative) is correlated with 0.26 and a p-value of 0.00.

tion in this area as the target market. Because of the widespread devastation in the area, large parts of the population were relocated to temporary camps inland, meaning that the physically untouched companies suffered from a loss in customers (FG-Fish-BNK). Nearly all of the business sectors in the region, except the retail sector, are more or less subject to seasonal fluctuations in their business activities. In the tourism sector, nearly 60 % of the companies are subject to seasonal changes, while in the agriculture and fishing sector, 40 % of the companies have to reduce their activities due to the monsoon weather changes. In Patong Beach, for example, the economic situation slowly relaxed in the next high season for tourist businesses starting in November, eleven months after the event, while other companies that were not dependent on seasonality felt the recovery earlier (FG-Tou-PB). This influence expresses itself in a longer recovery time in terms of sales volume, with a medium influence of an f^2 value of 0.03 (cp. Table 63). This influence also spreads out to the households, since most of the companies reduce their number of employees during the low season. The supply situation is ranked less important by the companies, but is nevertheless an essential part of the business activities (cp. Table 62). Similar to the situation for the physical recovery or the sales recovery, the consequences of indirect effects resulting from the interruption of lifelines showed the strongest effect on the duration of time needed to return to a stable equilibrium status (cp. Table 64). Showing an f^2 value of 0.13, it is obvious that the restoration of the infrastructure is the basis for the economic upswing of the area. This is also supported by the strong effect of the indicator 'Phang-Nga', which points to the fact that companies located in the extensively devastated areas of the provinces needed more time to recover due to the fact that large parts of the economic circle in this area were destroyed.

The factor 'backward linkages' contains the information about the number of business sectors acting as suppliers for this particular company, and therefore describes the company's status within the economic circle. With an effect size of 0.05 and a positive path coefficient, this factor points to the fact that companies depending on multiple business sectors to sustain their supply of raw materials and other supplies are likely to recover more slowly from the external shock. High backward linkages can especially be observed in the manufacturing sector, which depends on specialised raw materials and supplies, i.e. the fish processing companies in Ban Nam Khem, and the restaurant and hotel sector, which is partially dependent on supplies from the region in terms of food (FG-Hot-KL; FG-Ret-KL). Despite the restrictions resulting from the specialised supply needs, it can be found that companies that show higher backward linkages tended to use their own initiative to find new supply possibilities significantly more often. This fact is not correlated with the sales volume of the company and indicates that specialised companies, regardless of their size, are less prone to constraints in their recovery efforts. Due to the advanced economic level of the Phuket Province and the higher number of wholesalers on the island (mostly located around the capital of Phuket, which was not touched by the tsunami), several companies from the Phang-Nga Province rely on suppliers from the island, which resulted in virtually no additional interruptions of supply except those caused by interruptions of the road system (cp. Table 64).

Table 64: Recovery of supply (Own calculations based on TRAIT company survey)

	Path-coefficient	Indirect effect**	f^2	VIF	corr. R^2
Indirect recovery*	0.34	0.00	0.13	1.12	0.25
Backward-linkages*	0.19	0.00	0.05	1.12	
Sales location*	-0.20	0.00	0.05	1.14	
Location of supply*	-0.07	0.00	0.01	1.02	

* significant min 5 % error – TRAIT Company Survey – Durbin-Watson: 2.40
** Indirect effects are calculated as difference of total effects and direct effects (path-coefficient) – n = 199

8.2.3 Conclusion

In order to link all three sub-indices determining company vulnerability, i.e. sales, supply and physical recovery, a total vulnerability index was used. This encompasses the time the companies needed to return to a point where they considered themselves to be recovered. When normalising the three parts of exposure and using them as sub-index values for the explanation of the total vulnerability, the highest influence on the total recovery arises from the physical recovery of the company (0.41), the second strongest impact from the sales recovery (0.34) and the third strongest impact from the supply recovery (0.25).

As seen in the previous sub-vulnerability indices, the physical vulnerability of company property dominates the total vulnerability index with a strong f^2 value of 0.22. Here, the total effect of this factor is stronger than the direct link, which is represented by the path coefficient. This indirect effect within the model accounts for a standardised beta-coefficient of more than 0.12, which is mediated through the sales and supply sub-indices (cp. Table 65). Despite the fact that the upstream factors describing the location and the pre-tsunami asset volume show only a minimal effect size, which is no longer measurable, the amount of assets at least has a slightly enlarging effect on the total index. The fact that the 'location factor' shows no influence on the total vulnerability index in combination with the normalised beta-coefficients indicates that indirect effects account for a stronger influence on the total vulnerability than the physical impact itself. Therefore, companies which were completely untouched by the wave impact might show a similar recovery pattern to companies which were highly devastated by the waves. Here, it is obvious that the amount of damage caused by the tsunami shows no f^2 value and only a minimal total effect of 0.03 (cp. Table 65).

Table 65: Total vulnerability index for companies (Own calculations based on TRAIT company survey)

	Path-coefficient	Indirect effect**	f^2	VIF	corr. R^2
Physical recovery*	0.38	0.12	0.22	1.21	0.48
Recovery sales*	0.32	0.00	0.18	1.50	
Supply*	0.23	0.00	0.09	1.26	
Assets*		0.02	0.00		
Loan*		0.03	0.02		
Sales location*		-0.19	0.00		
Customer tourists*		0.03	0.00		
Direct effects*		0.03	0.00		
Education level*		-0.07	0.02		
Ext. help*		0.05	0.00		
Indirect recovery*		0.38	0.00		
Investor*		-0.04	0.00		
Location*		0.00	0.00		
Market trend*		0.05	0.00		
Value chain*		-0.06	0.00		
Supply location*		-0.02	0.00		
Backward-linkages*		0.04	0.00		
Age*		-0.09	0.00		

* significant min 5 % error – TRAIT Company Survey – Durbin-Watson: 1.84
** Indirect effects are calculated as difference of total effects and direct effects (path-coefficient) – n = 199

The external help provided by the government and NGOs, which can be counted as external resilience capacity, and which has mostly been targeted at the physical recovery of the companies, is only marginally measurable on the total company vulnerability index. Due to the limited number of companies and the low value of the goods given to the companies by the government or NGOs, the effect of this external help is not measurable in terms of the total recovery time for the companies. On the other hand, the low-interest loan programme initiated by the government has a small effect size on the total vulnerability index with an f^2 value of 0.02. Nevertheless, the dependency on external credit shows an extending influence on the total recovery time. On the one hand, this is surely due to the fact that only companies that suffered from physical damage received these kinds of credits, but on the other hand, the external credits extended the debt level of the companies, which were often already high, increasing their sensitivity to later shocks and decreasing their capacities for further adaptations to the challenges of winning back the customers and returning their business activities to normal. In the area around Khao Lak in particular, several businesses must have had large burdens from previous loans resulting from the strong expansions in the decade prior the tsunami (cp. Chapter 7.2.3).

Internal resilience capabilities were previously defined as characteristics of the company which allow them to cope with the outcome of the external shock more easily and to absorb or soften the effects of the shock on their business activities. This comprises all aspects of business organisation, sales or supply activities, which have already been discussed separately. In terms of the total vulnerability status outcome index, these internal resilience capabilities are mediated by the three pre-identified exposed elements of the company. Most of the effect size is directly concentrated on each of the sub-indices only, i.e. physical recovery, sales or supply, and is not measurable for the total recovery capability. The only difference can be found in the factor representing the educational level of the company. This factor directly influences the physical recovery time of the company, but it shows no effect size on the recovery of sales. Despite this, elements could be found in several interviews indicating that more highly educated managers seized opportunities such as founding additional commercial lines, for example construction companies, in order to benefit from the rebuilding process, or used other rebuilding-related business opportunities (FG-Tou-KL; TRAIT Company Survey). In Khao Lak and Patong Beach, the education level is significantly higher and is at least on the level of upper secondary school, while in the two other study regions, the education level mostly rests at the completion of lower primary school. Higher levels of education are not limited to the management level, but are also significantly correlated with the average education level of the employees.

9 Application of vulnerability model on the meso-level for the post-tsunami situation

9.1 Household vulnerability on the community level

In order to compare the study communities, the households' structural equation model has been adapted to all interviewed households in order to predict their vulnerability for their current situation using the reference scenario of the tsunami 2004 in terms of physical impact and the external help situation. Both input and output data for the structural equation model are z-transformed values, and therefore interpretations are only possible in relation to the study communities and in terms of significances of the mean values.

Comparing the four study villages to each other, it can be seen that the households in Ban Nam Khem recovered significantly more slowly than households in the other three study areas, in terms of both the physical rehabilitation to the pre-tsunami level and the recovery of household income. The significance test reveals that Ban Nam Khem shows the strongest restrictions in its recovery potential. The other three communities can be split up into two separate recovery groups: on the one hand, there is the community of Thai Muang, which was only slightly directly affected but severely indirectly affected, and shows similar recovery values to the municipality of Patong Beach, both in terms of direct and indirect recovery. On the other hand, there is the strongly affected community of Khao Lak, which can be placed between the slowly recovering community of Ban Nam Khem and the rapidly recovering communities of Thai Muang and Patong Beach (cp. Table 66).

Table 66: Comparison of vulnerability factors on the household level between the study communities (Own calculations based on TRAIT household survey)

		Ban Nam Khem	Thai Muang	Khao Lak	Patong Beach
Exposure	Assets	-0.19_a	-0.70_b	-0.20_a	0.31_c
Sensitivity	Location	-0.35_a	0.79_b	0.36_c	-0.04_d
	Tourist dependency	-0.14_a	-0.11_a	0.03_a	0.19_a
	External help from government	0.32_a	-0.19_b	-0.04_b	-0.20_b
	External help from NGO	0.20_a	-0.10_a	-0.06_a	-0.11_a
	Infrastructure effects	0.55_a	-0.45_b	0.12_c	-0.41_b
	Personal injuries	0.07_a	0.03_a	-0.01_a	-0.04_a
Resilience	Diversification	0.07_a	0.10_a	0.06_a	-0.17_a
	Work situation	-0.12_a	-0.16_a	0.05_a	0.18_a
	Own resources	0.22_a	-0.17_b	-0.07_a	-0.10_a
Recovery indices	Amount of damage	-0.10_a	-0.56_b	-0.35_c	-0.36_c
	Physical recovery	0.78_a	-0.46_b	-0.11_c	-0.50_b
	Recovery of income	0.46_a	-0.35_b	0.09_c	-0.35_b

Subscripts in the same column not sharing the same letters are significantly different at $p<0.05$ in the two-sided t-test - N = 323;

In terms of exposed assets, three classes and development stages can be observed among the four study communities. At the lower end of the scale, the small community of Thai Muang has the least exposed assets. Patong Beach, with its nearly 20,000 permanent residents, is at the upper end of the exposure scale because here, the economic development has made strong progress in recent decades and the households have been able to accumulate assets. Khao Lak and Ban Nam Khem show, on average, fewer accumulated exposed assets than Patong Beach due to the short increase in prosperity, in which only a small number of households have been able to participate so far. Comparing the location factor among the communities, it is obvious that in the current situation, Khao Lak and the indirectly affected community of Thai Muang show the highest location factors, making them less susceptible to a new tsunami. Following the widespread destruction during the tsunami of 2004, several households in Khao Lak were relocated to areas further inland in order to reduce the population's sensitivity (cp. Figure 34).

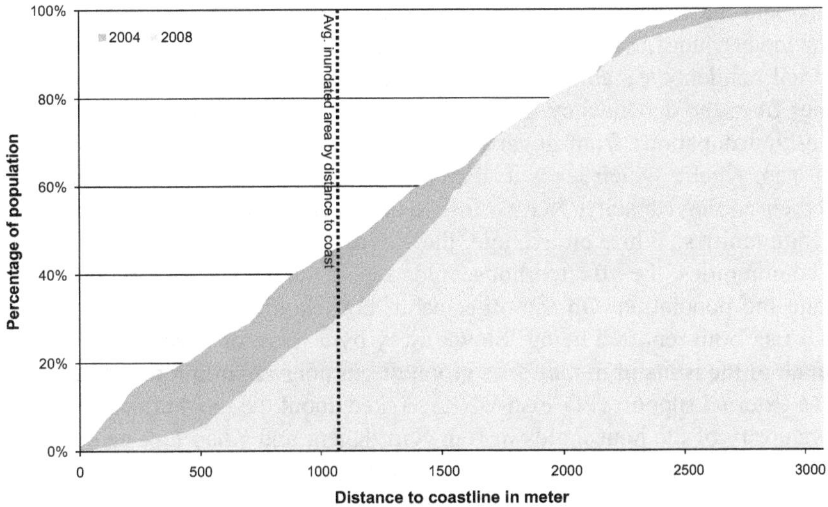

Figure 34: Location of population Khao Lak 2004 and 2008 (Own calculations based on TRAIT household survey and TRAIT MFC data)

This leads to an increase in safety, but also to displeasure among people working in fisheries and those who had lived close to the sea on public land before their houses were destroyed. Now, these groups of people have to accept a travelling time of up to 20 minutes to reach their fishing boats (FG-Fish-KL). Ban Nam Khem is still highly endangered with the lowest location factor, which indicates that the households in the community are mostly located in low-lying areas and close to the sea. Households in Patong Beach are, in general, further away and higher above sea level. For the households in Thai Muang, only marginal direct tsunami impacts need to be considered. Due to its safe location, the estimated amount of damage for Thai Muang is the lowest of the three other communities. Ban Nam Khem is still estimated to have been highly destroyed, but due to its low accumulation of assets, the amount of damage is still lower than in Khao Lak and Patong Beach. These two communities have a similar estimated damage appraisal due to the high accumulation of assets in Patong Beach and the continued large number of households which show a strong sensitivity towards the tsunami in Khao Lak.

As shown previously, the recovery potential of the households in terms of physical recovery capability is determined by the infrastructural constraints. Here, there are strong limitations on the households in Ban Nam Khem and Khao Lak due to the highly exposed lifelines along the coastline, and, as a consequence of this, the long duration of the recovery period. In contrast, Patong

Beach and the community of Thai Muang, which was only indirectly affected, show lower values, which allows the households to recover faster in terms of an external resilience capability. An extending influence on the recovery potential comes from the dependency on external recovery support in terms of financing or in-kind donations from governmental or non-governmental organisations in Ban Nam Khem, which leads to a slight extension of their recovery capability and their coping capacity. NGO efforts were spread significantly equally among all communities. While on average, the external support does not differ among the communities, the affected households report the uneven distribution of help among the population. On the other hand, households in Ban Nam Khem and Khao Lak both reported being 'blown away by a wave of generosity' in the aftermath of the tsunami in the focus group discussions regarding the effectiveness of the external support (FG-Fish-BNK). Asked about the most effective support, the majority of the households in Ban Nam Khem and Khao Lak named one of the several NGOs, which were mostly involved in building houses or providing occupational tools. Parallel to the experience from the quantitative household survey, the participants of the focus groups complained about the uneven distribution of help among the tsunami victims. While the government granted relatively even support for the tsunami victims, the NGOs reached only a limited number of victims (FG-Fish-BNK; FG-Fish-KL). This statement occurred in nearly all focus groups, where the participants observed "better and much more help from NGOs and international foundations than from the government" (FG-Fish-BNK). Similar to the observations from the modelling approach, the participants reported an extension of the recovery process as a consequence of the massive external support. This expressed itself in the various statements regarding the year after the tsunami, when some of the households stayed in temporary shelters and "did not have to do anything back then thanks to the consistent help offered in terms of money, food and utilities" (FG-Fish-KL). Similar statements were made in the community of Ban Nam Khem (FG-Inc-BNK). In the first period after the tsunami, the population experienced an insufficient management of resources and competences in the risk management, where as a consequence, the help provided by the NGOs sometimes had a "double standard and required no local participation", or people "who did not suffer a direct impact from the tsunami received lots of donated things and all types of compensation" (FG-Pla-KL). On the other hand, governmental support did not cover the actual losses, since the quota only reached, on average, not more than one-tenth of the total physical loss of the households (TRAIT Household Survey).

The possibility of the households to finance the recovery from their own resources, such as insurance claims, savings or from social networks, is classified as an internal resilience capability with a lowering influence. In this category, the households in Ban Nam Khem show the highest average values, which is due firstly to the higher damage than in the other study areas, but also to the absolute number of affected households using their own savings for the recovery process. Put in relation to the total number of affected households, the share is significantly smaller than in the other communities (TRAIT Household Survey). This supports the thesis that savings are an important factor for risk reduction and coping, but are mostly pursued by higher income level households (HORWICH 2000:527). Additionally, the share of insurance for covering risk from natural hazards is strongly linked to the income level and grows with increasing income (HORWICH 2000:528). In the surveys, it was found that several households bought insurance directly after the disaster or participated in community savings groups, but cancelled these risk transfer possibilities soon after due to budget constraints (TRAIT Household Survey, FG-Fish-BNK). Despite this, it can be assumed that in the case of a more serious devastation in the other areas, the households are still able to raise their own resources for the recovery process if a certain point of necessity is reached.

The last factor found to influence the physical recovery capabilities from the internal resilience perspective is the personal injuries of household members, including killed or wounded household members. Since the tsunami occurred during the daytime, many people were not in their houses, and that location had only limited influence on the number of personal incidents. As a result, the number of people killed or wounded does not vary significantly between the study communities. Besides the psychological influence on the household, this also had a financial impact resulting from the costs for the funeral and the loss of a worker to the household (THANAWOOD et al. 2006:213; FG-Inc-BNK).

Taking a look at the indirect effects, i.e. the loss or reduction of household income resulting from the disaster, it can be seen that the least time required can be found in Thai Muang and Patong Beach, which were only indirectly or slightly affected. At the same time, Ban Nam Khem and Khao Lak show significantly higher values. Therefore, the households' adaptation strategies and their access to livelihood assets among these two groups of study areas need to be investigated. For the affected households in the region, the indicators related to the household's 'work situation' showed a particularly significant influence. These indicators do not differ significantly between the three study communities, indicating mostly internal disparities within the communities. The degree of diversification in particular, measured by the Shannon index on the household level, directly influences the time of indirect effects. This index value does not vary significantly between the three study communities, but mostly within the population of each community. The highest scores were reached in Thai Muang. The reason for the differences among the villages is the domination of single business sectors in each of the villages. Around one quarter of the households' income is generated directly in two dominant business sectors, and also in various downstream industries depending on the input of agriculture and fishing activities. Due to the climate conditions in southern Thailand, small-scale coastal fishing activities are only possible outside the monsoon seasons from May to October. As a consequence, many of the small-scale fishing families owning only a traditional long-tail boat of around 10 m in length depend on part-time jobs or additional income possibilities during the monsoon season. These mostly consist of jobs on plantations or low-skilled jobs such as motorcycle taxi driver or garbage collector (FG-Inc-BNK, FG-Fish-KL). Despite the significant negative correlation between the level of household income and the share of income generated from agriculture and fishing, a strong positive influence on the recovery time can be seen, which allows the households in Thai Muang a faster recovery, and therefore a reduced vulnerability. The non-significantly differing share of income generated in tourism-related businesses in Thai Muang can be ascribed to the commuter linkages to Khao Lak, and therefore the spreading of indirect effects from there towards the surrounding areas. In Thai Muang, the household members mostly stayed in their old occupation and waited for the recovery of their business activities. In the other communities, the household members needed to switch their occupation after the event due to the extended business interruptions. Most people participated in programmes developed by the government targeting the unemployed. These job programmes provided mostly low-skilled manual services, such as garbage collecting or water delivery. On average, the district government paid minimal wages varying between 50 and around 100 € for these services, which implies a strong reduction in income for the affected households (FG-Fish-KL). As seen in the quantitative model, this reduced income had a strong extending effect on the recovery time of the households. These jobs were performed by the local population for a period of between six months and two years after the tsunami, when the local economy returned to its full, normal power. Besides these service jobs, the government agencies used paid further education as a tool to reduce unemployment. In groups, skills were taught such as handicraft, batik printing, basket making or the cultivation of mushrooms, or services such as baking, hair dressing, and others which were intended to empower the population to run their own business. In the long run, most of the participants in these groups returned to their previous occupations or to another form of employment, and only a limited number of these shops could be observed in Khao Lak or Ban Nam Khem (FG-Inc-KL; Field mapping). Despite its strong tourism dependency and the high indirect effects in absolute terms, the households in Patong Beach can be counted as having recovered rapidly in terms of loss of income, which is due to the strong marketing efforts for the area (cp. Chapter 6.4.4).

9.2 Company vulnerability on the community level

In order to assess the vulnerability of the study communities for the current situation, the company structural equation model was adapted to all interviewed companies using the reference scenario of the 2004 tsunami in terms of physical impact and the external help situation. The risk analysis in terms of physical damage is based on the remote sensing results in combination with the elements of the structural equation model in order to predict the amount of damage for the current situation.

In terms of exposed elements, Patong Beach shows the highest values in terms of companies' productive assets. Due to the high agglomeration of companies in the community, these values differ significantly from the other communities. Since all assets in the city are included, it can only be seen as a theoretical value, which has to be set in contrast to the values that are located inside the potential inundation zone. In combination with the location factor, Patong Beach reaches an amount of damage which is minimal compared to the amount of exposed assets. In Khao Lak, the total amount of damage is slightly above the average, and in Ban Nam Khem, with its relatively small amount of exposed assets, it reaches only above-average values (cp. Table 67).

Table 67: Company vulnerability aggregated to the community level (Own calculations based on TRAIT company survey)

		Ban Nam Khem	**Thai Muang**	**Khao Lak**	**Patong Beach**
Exposure	Assets	$-0.11_{a,b}$	$-0.11_{a,b}$	-0.07_{a}	0.11_{b}
Sensitivity	Location	-0.33_{a}	0.32_{b}	-0.20_{c}	-0.12_{d}
	Tourism dependency	-0.71_{a}	$-0.26_{a,b}$	0.13_{b}	$0.14_{b,c}$
	Credit	-0.12_{a}	-0.13_{a}	0.12_{a}	-0.02_{a}
	External help	0.20_{a}	0.01_{a}	0.12_{a}	-0.09_{a}
	Infrastructure effects	0.24_{a}	-0.24_{a}	0.12_{a}	-0.07_{a}
	Seasonality of market	-0.24_{a}	$-0.06_{a,b}$	0.49_{b}	-0.13_{a}
	Location of sales	1.18_{a}	1.18_{a}	1.18_{a}	-0.85_{b}
	Location of supply	1.22_{a}	-0.11_{b}	0.59_{c}	-0.49_{b}
Resilience	Age	0.52_{a}	$-0.08_{a,b}$	-0.26_{b}	$-0.01_{b,c}$
	Education	-0.57_{a}	-1.00_{a}	0.02_{b}	0.24_{b}
	Integration in value chain	-0.39_{a}	$0.71_{a,b}$	0.18_{b}	$-0.06_{b,c}$
	Backward linkages	-0.51_{a}	$-0.10_{a,b}$	0.17_{b}	$0.06_{b,c}$
	External investor	-0.09_{a}	-0.06_{a}	0.28_{a}	-0.08_{a}
Recovery indices	Physical recovery	0.84_{a}	-0.37_{b}	$0.21_{a,b}$	-0.23_{b}
	Amount of damage	0.13_{a}	-0.56_{b}	0.02_{c}	-0.26_{c}
	Recovery of sales	0.58_{a}	-0.35_{b}	0.09_{b}	-0.12_{b}
	Recovery of supply	0.27_{a}	-0.12_{a}	-0.01_{a}	-0.04_{a}
	Total recovery	0.75_{a}	$0.01_{b,c}$	0.24_{c}	-0.23_{b}

Subscripts in the same column not sharing the same letters are significantly different at $p<0.05$ in the two-sided t-test - N=295

The location factor in this analysis is aggregated above all companies in the community. Despite this, the companies in Ban Nam Khem are nearly all highly endangered by a tsunami impact with the characteristics of the one in 2004. Despite its widely spread area, the companies in Khao Lak have a significantly similar location factor, which is due the fact that they are mostly located along the coastline, and, because of their tourism orientation, were not relocated inland. The location factor in Patong Beach for all companies is slightly lower than the one for Ban Nam Khem and Khao Lak due to the community's extension up through the mountain area into higher ground, which reduces the overall location index for the companies here. In Thai Muang, no direct impact on the company's property is likely due to the protected location of the area.

As shown in the previous chapter, the amount of total damage in company property plays only a secondary role in the recovery process of a company. This process is almost completely dominated by the internal and external resilience potential of the company. In Ban Nam Khem, these capacities are less represented, which can be seen by looking at the recovery time for physical vulnerability, which is higher than average, but on the community level not significantly different to the other study communities. On the company level, the dependency on external help shows a strong extending influence on the physical recovery time. On the level of the communities, no significant differences can be found in terms of the donation of goods or low-interest credits. The external influencing factor of infrastructural effects is not significantly different either when comparing its influence on the community level. Significant differences on the community level can be found in terms of internal resilience capabilities. The education of the company's owner is marginal in Ban Nam Khem and Thai Muang in particular, and therefore not able to compensate for the lack of other resilience capabilities. These capabilities are strongly developed in the municipality of Patong Beach and, to a lesser degree, also in Khao Lak. Companies in Ban Nam Khem could also strongly benefit from the experience of their owners, represented by the indicator of company age. Here, a significant difference can be found, indicating that the companies in Khao Lak have the least experience and therefore the least benefit from the reducing influence of company age.

In terms of the recovery from the loss of sales, Thai Muang and Patong Beach show the fastest recovery time and are also significantly similar in this variable. Thai Muang, in this case, can benefit from the low dependency on income from tourism. Only indirect effects resulting from the interrelations to companies and consumers in the largely devastated coastal areas of Phang-Nga influence the business activities of this area. In Patong Beach, on the other hand, the fast recovery of the companies in terms of sales was fostered by the economic surroundings, which had been less affected in direct terms. However, the businesses in Patong Beach show a low integration in the value chain, which deprives them of the opportunity of switching to other markets. This opportunity was available to the companies in Thai Muang. Ban Nam Khem, on the other hand, showed the least recovery potential in terms of sales, which is mostly due to the strong integration and focus on the regional markets in Phang-Nga. The mostly agriculture and fishing companies in Ban Nam Khem are subject to seasonal market changes, which made it difficult for them to recover quickly. Additionally, the integration of all companies into the value chain in Ban Nam Khem is significantly lower than in all other regions due to the high number of small retail shops targeting the local population. In terms of supply interruptions, the communities show no significant differences. This fact may result from their overall strong orientation towards suppliers located in the district capitals. Differences in this factor can only be revealed when looking at the micro-level and the company itself.

In conclusion, the total vulnerability index can be investigated for the four differently positioned study areas. As seen in the previous chapter, this index represents the entire group of observed factors and indicators and their interrelations. It combines the three parts of business recovery elements, representing the vulnerable status and its coping capacity for recovering or returning to a stable status. On the aggregated community level, three vulnerability groups can be observed. Firstly, there is the highly vulnerable community of Ban Nam Khem, whose total recovery index points to a highly vulnerable status not comparable to the other communities. This results from its outstanding damage ratio. Due to this fact and the low values for the elements determining the physical recovery capability, it also suffers from long-lasting indirect effects in terms of sales reductions, which cause the overall highly vulnerable situation in the community. A second vulnerability group can be found in the two study areas of Thai Muang and Khao Lak, whose vulnerability index is significantly similar. Whereas the degree of destruction in Khao Lak can be compared more to the situation in Ban Nam Khem, its physical recovery capabilities arising from the various determining factors help to achieve a faster recovery and better coping capability. This allows a much faster overall recovery from the effects of the disaster. The small community of Thai Muang is coupled in economic terms to the development in Khao Lak, and therefore also in its vulnerability depending on the "lighthouse" of Khao Lak, despite the marginal direct effects. The recovery of sales in particular determines the vulnerability of the companies in Thai Muang, which is dependent on the situation in Khao Lak. In terms of the average damage to company property, Khao Lak is comparable to Ban Nam Khem, but its companies have more coping capacity available for dealing with the outcome, meaning that the physical recovery capacity is similar to that of the less affected areas such as Patong Beach. The sales situation is also comparable, but still slightly longer, meaning that overall, Khao Lak's vulnerability can be placed between Ban Nam Khem and the least affected municipality of Patong Beach. In comparison to the other communities, Patong Beach had the highest coping capacity for dealing with the physical damage, since the companies' physical recovery time is comparable to that of Thai Muang. In terms of the recovery of sales, Patong Beach had to struggle a little bit longer than Thai Muang. This is most likely the influence of its tourism dependency and its lower integration into the value chain, which allowed the companies in Thai Muang to look for other customer groups or markets, while the companies in Patong Beach had to wait for the return of the tourists.

10 Conclusion

10.1 Summarising the empirical findings

Ideally, vulnerability assessment aims at a comprehensive and holistic assessment of the status of a specific sphere and the processes which are started by the impact of an external shock. Theoretically based and proofed in the empirical evaluation, it could be found that the vulnerable status and the processes which follow are composed of the three system perspectives of exposure, sensitivity and resilience, encompassing internal and external capabilities and characteristics. In order to achieve a most holistic vulnerability assessment, the derived structural equation model integrates elements of vulnerability research from the different scientific disciplines into a single model. Elements of vulnerability assessment from economics, social sciences and physical risk assessment were integrated. This holistic vulnerability assessment leads to a differentiated outcome of the vulnerable status of households and companies on the micro-level. In the following section, the guiding questions set up in the 'Introduction' will be taken up and conclusively answered.

E_1 Which differences can be seen between the study areas in terms of directly and indirectly exposed assets?

In order to estimate the coping capacity of households and companies towards the direct and indirect effects of natural hazards, these need to be quantified. First, a regression-based procedure was developed to estimate the total amount of assets per household based on remote sensing indicators. In a second step, the thus specified parts of the structural equation model were applied to the households and companies. The results showed that the model could reproduce the actual damage sufficiently and allowed an interpolation of the survey results to other non-surveyed households and companies, as well as an application of this model to the current situation, with the support of remote sensing pictures. The outcome reveals that the damages were very unevenly distributed among the communities because of the varying size and settlement density, but also because of the different development levels when investigating the household income, with Khao Lak and Patong Beach in the one category and Thai Muang and Ban Nam Khem in the other. The agglomeration of businesses' assets also shows different levels due to the existence of more medium and large companies in Khao Lak and Patong Beach, and the nearly exclusive focus on small and micro-businesses in Ban Nam Khem and Thai Muang. Therefore, in terms of absolute accumulation of household assets, Ban Nam Khem with around 5.5 million

€ and Thai Muang with around 0.37 million € made up one group, while the total accumulation of household assets in Khao Lak and Patong Beach, with around 18 million and 55 million € respectively, is on another, higher level. When comparing the estimated damage of the 2004 tsunami, it can be seen that Ban Nam Khem and Patong Beach are in the same class for the total damages. The reason for this is that in Ban Nam Khem, nearly all households are located in tsunami-endangered areas, while in Patong, only 20 % of the households live that close to the sea. In Khao Lak, the majority of households are located inland, which produces a similar situation to Thai Muang, where virtually no damage to the household location-bound assets occurred. In terms of destroyed company assets, the tourism orientation of the local economy and the concentration of assets along the coastline resulted in relatively high damages in both Patong Beach and Khao Lak of an estimated 30 and 90 million € respectively. Despite its economy being completely located within the 2004 inundation zone, the five million € worth of damage in Ban Nam Khem is relatively low compared to the other communities. For the current situation, a noticeable change could only be found for the households in Khao Lak due to massive relocation measures. For Ban Nam Khem, the majority of the households have remained in the endangered zones. In terms of companies, no relocations and therefore no changes in the exposure or the estimated damage could be found.

By observing the recovery times of the households and companies in the study areas, it can be seen that these are significantly different and not related to the total amount of damage. Therefore, the coping capacity had to be investigated.

E_2 What are the interdependencies between the elements of vulnerability? Which elements reduce vulnerability? Which increase the vulnerability of households? How are the relations quantified?

In order to capture the interdependencies within the elements and perspectives of vulnerability in a structural equation model, the pre-tsunami status, the impact and the recovery process afterwards were modelled following a causal and chronology chain. The physical impact for household property could be remodelled, benefiting from the progress in the strand of risk analysis vulnerability by combining a location factor and the pre-tsunami assets volume. For both households and companies, a sufficient corrected R^2 could be achieved, which supports the theoretical assumptions and the representing indicators. It was hypothesised for both households and companies that the amount of physical damage caused by the occurrence of the natural hazard influences the vulnerability proxy of overall recovery time. In the vulnerability situation of the households, no such influence could be identified, which points to the fact that this part of vulnerability could be completely mediated by other factors resulting from the resilience perspective. In terms of companies, the amount of damage could be traced back throughout the whole process of recovery and still showed a significant influence, which indicates that the coherences in company vulnerability are not completely parallel to the household processes.

The physical exposure and sensitivity is a crucial point for the companies' vulnerability, whereas for the households' vulnerability, this hypothesis must be rejected.

For the household vulnerability, it could be identified in step-by-step modelling that the level of damage to household property is completely mediated by the financial capabilities on which the household can rely to restore its loss. The vulnerability level is thus strongly determined by the question of whether the household needs to rely on external support or can fall back on support from social networks or on its own savings. Dependency on external support increased the vulnerability proxy compared to support from social networks or a household's own savings because of its slow and bureaucratic disbursement, despite all efforts. By observing the indirect effects within the structural equation model and the differences among the communities, it can be seen that the availability of social networks or savings is not dependent on the overall income level, since the communities with a high income level show less reliance with this kind of compensation source, i.e. comparing Ban Nam Khem and Khao Lak.

Looking at the vulnerability and recovery process of companies, no direct influence of the amount of damage on the recovery time from the physical impact could be found either. But in contrast to the household model, a significant indirect effect moderated through the amount of the loan to the company can be observed, which clearly indicates the government's adequate reaction to the level of devastation. External help in the form of donations of goods only played a relevant role for micro-businesses and cannot be linked to the amount of damage. A crucial point for the implementation of external aid and support programmes would be non-bureaucratic, equal distribution and swift handling. This point was not fulfilled after the tsunami of 2004, since a limiting effect for the economy could be found in comparison to others who were able to finance this process from their own capacities. Previously, a similar pattern had also been observed in the recovery efforts after the Kobe earthquake or Hurricane Andrew in the United States (HORWICH 2000; WEBB et al. 2002).

For households, external and internal monetary capabilities represent the strongest influence in their physical recovery whereas the financing is just a secondary issue for companies and covered by other external issues like disrupted infrastructure.

Furthermore, indirect effects were included in the models for the economic sphere. On the company level, these indirect effects express themselves in terms of influencing the companies' monetary and commodity flows to suppliers and customers. On the household level, which, in the economic cycle, represents the final demand, income reductions were included, while shortages in supply were left out due to the easily interchangeable supply in this locally restricted disaster. Overall, these elements show an important influence on the chosen vulnerability indices. These factors themselves are subject to elements deriving from each of the three perspectives of vulnerability. In terms of the households, these result from elements describing the employment of the household members, i.e. diversification of income, education and income level, or income from the region's dominant and especially susceptible business sector. Due to their direction of influence, it can be concluded that these 'capitals' or 'assets' such as income diversification, education and income level increase the coping capacity of the household. On the other hand, a higher share of income from tourism increases the possibility and duration of suffering from indirect effects, and should therefore be counted as an element defining the sensitivity of a household towards indirect effects. From his study on the influence of economic diversification on the recovery process following earthquakes in countries in different economic development stages, RAMCHARAN (2005:15) reasons that economic specialisation can extend the economic impact of earthquakes by inducing indirect influences on the income situation of households and companies, especially when no insurance coverage is available to compensate for the losses. This finding from the households can also be verified for companies, where tourism-dependent companies showed an extended and more likely loss of sales than other companies, and were also prone to seasonal changes in their business activities resulting from the changes in tourist numbers over the year. In need of special mention is the resilience element of education, which is represented in the models by the average education of the working household members or by the education of management personnel, and has a strong increasing impact on the overall coping capacity in both models, and therefore reduces vulnerability. This asset of households and companies can strongly determine the recovery process by offering possibilities in terms of changing the workplace, organising external help or accessing new market opportunities.

Internal resilience capabilities of households and companies can compensate for physical damage and reduce the influence of indirect effects.

E_3 Which household and company groups can be identified as more vulnerable to natural hazards?

On the micro-level in particular, interactions determining vulnerability can be observed very clearly by studying the direct and indirect influences on the chosen vulnerability index. From these rankings, it can be determined as to which characteristics of households and companies make them especially vulnerable to hazards like the tsunami of 2004 (cp. Table 68; Table 69).

On the household level, it is obvious that the degree of damage does not show a significant influence, indicating that the impact is completely mediated by coping capabilities. It can thus be concluded that the accumulation of assets alone does not influence vulnerability, either positively or negatively. Only in combination with resilience capabilities, which arise from the household's access to tangible and intangible assets, can vulnerable groups be identified. These assets can, in theoretical terms, be traced back to the approach of SEN and the sustainable livelihood theory. One's own capabilities in terms of financing the recovery process show the strongest influence on the recovery capacity, which underlines the influence of the ability to have a possibility for risk transfer, either through an insurance company or in informal ways using social networks such as family or friends. The 'own resources' category most often used in the study regions was that of risk transfer to social networks, which can be seen as fulfilling the function of formal insurance companies in emerging countries such as Thailand. Here, households which are socially integrated into the communities benefit especially. This is particularly the case in communities which are not defined by rapid growth and permanent inward migration. Having a significant but 'weak' influence on the occurrence and duration of loss of income, the income diversification and the education and income level show only a marginal explanatory influence on the final vulnerability index (cp. Table 68). Here, it is obvious that the households with the highest education level are employed in the service sectors, which are dependent on tourism-induced cash inflows, while people employed in agriculture and fishing are significantly less educated. A similar pattern is observable for the level of income diversification on the household level.

Table 68: Ranking of elements determining household vulnerability (Own calculations based on TRAIT household survey)

Element of vulnerability	Rank	Effect
Support from NGO	1	Increasing
Disruption of infrastructure	2	Increasing
Usage of own resources	3	Lowering
Personal injuries	3	Increasing
Loss or reduction of income	4	Increasing
Support from government	5	Increasing
Income diversification	6	Lowering
Education and income level	6	Lowering

* Damage level, Location, Income from tourism and Amount of assets show no significant influence - Elements are ranked according to their f^2 explanatory influence and secondly ranked after their total effect

The outcome of the model indicates that households depending on external support have the highest vulnerability, which manifests itself in the extended recovery for household depending on governmental or NGO support. Due to the nature of this study as a mainly quantitative vulnerability assessment, the model leaves it unclear as to whether the outcome of the NGO and government help results in an improved state relative to the situation prior to the event. This can only be answered by including the outcome of the focus groups and the qualitative parts of the surveys. While in the focus group discussions, a consensus had been reached that the support given was uneven and highly bureaucratic, in some areas the target of 'building back better' had been achieved. In Ban Nam Khem an upgrading of the overall building structure had been achieved with particular success. In some cases, the influence of the external help also lead to constraints in the everyday life of the population due to relocation of fishermen inland or low-quality buildings which did not meet the requirements of the local population. These relocations of fisherman into the hinterland have been made possible by missing land titles or houses build on public land, which have been destroyed and a rebuilding had been permitted (PETCHA-EM 2008). Another crucial point is the determination of the starting point of the household's recovery process by the effects resulting from infrastructural constraints. When starting a relief and recovery process, the rebuilding of infrastructure needs to have the highest priority.

On the company side, the choice of indicator 'return to recovered business activity' leads to a differing ranking of vulnerability elements. The proxy indicators for the recovery of company assets, sales and supply dominate the overall vulnerability proxy. Nevertheless, the access to tangible and intangible assets can also be seen in this case (cp. Table 69).

Table 69: Ranking of elements determining company vulnerability (Own calculations based on TRAIT company survey)

Element of vulnerability	**Rank**	**Effect**
Physical recovery time	1	Increasing
Recovery of sales	2	Increasing
Recovery of supply	3	Increasing
Education level	4	Lowering
Access to credit	4	Increasing
Recovery of infrastructure	7	Increasing
Degree to which primary sales location is affected	8	Increasing
Age of company	9	Lowering
Level of integration value chain	10	Lowering
Amount of external help	11	Increasing
Seasonality of business	12	Increasing
External investor	13	Lowering
Diversification of backward linkages	14	Lowering
Percentage of sales to tourists	15	Increasing
Amount of damage	16	Increasing
Amount of assets	17	Increasing
Degree to which primary supply location is affected	18	Increasing

* Location of company has no measurable influence - Elements are ranked according to their f^2 explanatory influence and secondly ranked according to their total effect

The education level stands out especially, showing the highest influence on the total vulnerability index. Here, a compensation element for the tourism industry can be found, since their management is significantly better educated than that of the other business sectors. Together with the age of a company, it can be seen as an element representing adaptation capacity to the new situation, as well as experience and networks on which the business can rely. Besides this, the availability of an external investor able to provide additional funding helps companies to reduce their vulnerability. This results in an informal extension of the capital stock of the company compared to other external support, and can also lead to positive linkage effects on other companies which might benefit from a faster recovery of their business partner (REVILLA DIEZ 1998:201). This also expresses itself in the positive effects for companies that are strongly integrated into the economic cycle and benefit from this through a faster recovery of their sales activities, which is an essential element of their overall vulnerability. For this element, companies from the business sectors of agriculture and fishing in particular as well as manufacturing companies show the highest values, meaning that their resilience potential is increased by these characteristics.

First of all, a negative influence on the overall vulnerability arises from the dependency on external help such as low-interest loans or donations, which is parallel to the household vulnerability model. Together with the effects resulting from infrastructural constraints, the control variables for the primary sales, and in limited terms the supplier locations, also indicate that the degree to which the surroundings are affected plays a significant role. This leads to the interpretation that a critical point in the recovery process, besides the re-establishment of the infrastructure, is the overall stimulation of the economy in the affected areas by governmental investment or other economic stimulus packages. Overall, the dependency on the tourism business only has a very limited influence on the overall vulnerability. This fact results from the differences of the two tourism-dominated communities. While in Patong Beach the tourism recovered very fast, in Khao Lak, the stigmatisation of the place and the seasonality of the business determined the recovery of sales. Therefore, the element might be reduced in its influence by the seasonal influences of this business sector.

E_4 Which differences can be seen between the study sites?

Aggregating these outcomes to the community level under the assumption of the tsunami 2004 scenario for the socio-economic setting of 2009 reveals the vulnerability on the meso-level. Due to the fact that the damage level has no significant influence on the total vulnerability of households, and only a marginal influence on the vulnerability of companies, particular attention must be paid to the resilience elements within the communities. Ban Nam Khem shows the highest vulnerability in terms of both direct and indirect influences due to its dependency on external help and its low education and income level. Khao Lak can be ranked second in terms of both direct and indirect impacts. Here, the vulnerability of the household is determined by the dependency on income from tourism and the resulting seasonal changes as well as the long stigmatisation of the location in the aftermath of the tragedy. Furthermore, the households showed a strong dependency on external help from NGOs and government agencies, and the households in the area also had the lowest rate of their own resources for restoration purposes. This may be caused by the strong inward migration into the community, which could result in a less strongly formed social network, which functions as a risk transfer tool. This lack of social network is insufficiently compensated for by the elements of income diversification, education and income level. A special role is played by the two communities of Patong Beach and Thai Muang, since both show significantly similar vulnerability indices on the household level. While no damage can be modelled in Thai Muang, households in Patong Beach have a significantly similar damage level to the situation in Khao Lak after the relocation process. This damage level can be compensated for by the resilience factor of education and income level, the extremely low dependency on external help and the virtually untouched surroundings, which enables a more immediate begin to the reconstruction process. In Thai Muang, a slight aid dependency can be found, and some of the households are also burdened by personal injuries of family members working in nearby communities close to the sea or as fishermen.

The situation on the company level is not completely similar to that of the households due to the different levels of integration in the value chain, their integration in local business processes and their professionalisation among the communities, and the slight but significant influence of damage level on their total vulnerability. Similarly, Ban Nam Khem shows the highest vulnerability resulting from the largest amount of damage and the severe lack of all resilience capabilities that would allow the businesses to cope with the situation. Furthermore, the businesses are least integrated into the value chain and mostly focused on the surrounding area, which would be completely devastated by a tsunami comparable to the 2004 scenario. Here, no improvement of the situation worth mentioning can be found. The situation in Khao Lak must be judged similarly, where the second highest damage level can be modelled due to the ongoing concentration of the businesses in the areas along the coastline. Despite the considerable physical impact, the area is also prone to longer lasting indirect effects resulting from the strong focus on services demanding the physical presence of the customer, which makes it necessary to move the point of sale in order to reach new customer bases, and also produces the highest seasonality dependency. For the tourism-linked businesses in particular, the absence of tourism for one season results in the loss of income for a whole year, and therefore considerably reduces the financial capital base of the companies and pushes them into a dependency on loans, which leads to an increased vulnerability. A special importance can be attributed to the economic setting in the study area of Thai Muang, since the businesses and households are not located in a tsunami area, and therefore no physical influence can be modelled. Despite this fact, the businesses there are ranked above Patong Beach in terms of their vulnerability, which has a minimal physical impact. Thai Muang's vulnerability score arises solely from the indirect impact and its coping and resilience capabilities, which are the least pronounced in most terms. The dependency on external help in the form of donated occupational goods is particularly high, as are internal resilience capabilities in terms of education level, which means that flexibility is lacking, which expressed itself, for example, in the lowest share of persons changing their workplace in the aftermath of the 2004 tsunami. The situation in Patong Beach is almost completely opposed to that of Thai Muang. Because here, companies are also offering services which demand customer contact to tourists and are almost completely dependent on tourism, the time of stigmatisation after the hazard, which was longer in Khao Lak than in Patong Beach, induces indirect effects to the complete regional economic circle, and therefore a dependency on loans due to the reduced sales. Due to the functioning infrastructure and other internal coping capacities, the vulnerability level of the municipality is still low. Besides the significant differences among the communities, a clearer and more precise picture can be observed on the micro-level, which is more likely to target deficits and needs in order to reduce vulnerability and increase resilience capabilities.

10.2 Political implications

Overall, the region experienced a strong media interest and a tremendous inflow of foreign aid in terms of the donation of goods and financial support. Additionally, efforts were made to improve the early warning and risk management and to increase the coping capacities of the local population. Furthermore, marketing efforts were made to re-establish the region's tourism potential and achieve a repositioning of parts of the study area on the global tourism market. In the following section, these efforts will be discussed in combination with the outcome and experiences from the modelling.

P_1 What kinds of adaptation measures have been taken on the personal level? Which measures have been taken by governmental and non-governmental organisations?

P_2 What implications do the findings of this work have for the endangered households and companies situated in potentially tsunami-prone areas?

On the household level, only limited strategies could be identified for mitigating future natural hazards. Due to the fact that the 2004 tsunami was rated as a once in a lifetime event, which statistically occurs only once every 550 to 700 years, the local population lost interest in developing further risk adaptation strategies step by step. While in the years after the tsunami several households procured insurance especially for the impact of natural hazards, several of them reported having cancelled their policy again. On the other hand, the participation rate in 'community-savings groups', especially in Ban Nam Khem and Khao Lak, is a common phenomenon and might increase the households' coping capacity significantly, as seen in the model from the item 'own resources'. Here, it is crucial to maintain the awareness of the necessity of this savings group in order to achieve a long-term increase in resilience. Occupational changes in the aftermath of the tsunami could be identified only in limited terms, mostly from persons previously employed in the agriculture and fishing sector changing to tourism due to the psychological consequences and/or the possibility of achieving a higher income. Confronting this fact with the outcome of the household vulnerability model, this occupational change needs to be deemed risky due to the strong seasonal changes in this business which might have a stronger influence than the additional income. Additionally, it decreases the region's diversification in business and leads to an overall stronger dependency on external inflow of income from tourism.

Several households interviewed showed a more fatalistic attitude when asked about their preparation for future hazards, and had completely neglected further preparations. It was not the focus of this study to identify the reasons for this attitude, though it may result from the psychological consequences of the 2004 disaster or a specific religious background, but it needs to be evaluated as a major constraint in risk preparation and adaptation enhancement. Additionally, the participation rate in regular tsunami evacuation drills has been shrinking over recent years. Despite this, the prognostic symptoms of tsunamis, for example the drawing back of the ocean, are known among the population and are also regularly taught at local schools. Furthermore, the regular drills and warning signal tests maintain the public awareness of the disaster, meaning that the number of dead and injured would be drastically reduced. Besides the awareness and the knowledge about the possible occurrence of tsunamis in that particular area, some households also relocated their residences further inland. This had happened only in very limited numbers on citizens' own initiative, and was mostly a consequence of the move into a new house donated by governmental or non-governmental agencies. In several cases, these relocation measures broke old neighbourhood networks and extended commuting distances to the workplaces, especially for people working in the fishing sector. Except for the micro-enterprises targeting neighbourhood customers in residential areas, most of the businesses did not relocate further inland, and no significant overall change in their 'location factor' can be found. Despite this, formal and informal building codes were at least established in Khao Lak, requiring the hotels to offer sufficient evacuation space for their guests from the coast-side bungalows. In the other communities, no such building codes have been established, though the companies trust the early warning system and, in some cases, are directly linked to the system. Regular drills with employees are also conducted. Only a small number of companies implemented mitigation strategies to enhance their resilience capabilities, such as evaluating new markets or starting sideline business activities. Furthermore, the current situation of the companies has an increased vulnerability due to the high loans needed to cope with the 2004 tsunami.

As indicated above, the majority of risk management and adaptation strategies applied to the region's economic system resulted from the initiative of governmental and non-governmental organisations as top-down approaches. From the household and company perspectives, the first aid and the cleaning and recovery of infrastructure, for example infrastructural lifelines, was effectively handled by the authorities. As seen in the outcome of the model, constraints in the recovery activities mostly arise from programmes developed in the period after the first aid phase. Both companies and households, as well as the findings of this study, criticise the slow imbursement rate of the donated money, the selective funding and the bureaucratic procedure for applying for loans or the donation of goods. In order to increase the internal financial capabilities of the companies and households, incentives for savings, reserve funds and insurance policies need to be provided for both economic elements. This problem has only been partially addressed by introducing community savings groups or the low-interest loans. Due to the limited term of the loans and the appearance of further economic problems resulting from global economic effects and political instability in Thailand, the companies have struggled to repay the loans and can be deemed even more vulnerable now. Another recovery initiative, which can also be seen as a mitigation strategy, is the occupational training programmes conducted to compensate for the loss of income after the tsunami and to increase households' access to 'assets' such as education and occupational possibilities. Another remarkable long-term mitigation strategy has been the implementation of a marketing campaign for the area of Khao Lak. Before the tsunami, the region experienced a constant growth and intensification of the tourism business, leading to a developed tourism area comparable to Phuket Island. Triggered by the tsunami impact, this strategy was revised and an ecologically based extensive tourism strategy was developed, as it could be identified that this customer group demonstrates a high spending power and solid return rates. This could increase the coping capacity of the region if the strategy helps to increase the capital base of the companies, and if the customers are more loyal and return faster after a natural disaster.

Overall, it can be concluded that the vulnerability of the region could be reduced by the measures implemented after the tsunami, but several necessities for further improvements nevertheless remain unaddressed. The highest priority goes to the promotion and improvement of the economic system's internal capacities, such as the diversification of business structures and the extension of the financial base. Parallel to this, the access to assets such as education and occupational opportunities needs to be improved.

10.3 Domains requiring further research

As shown in the previous discussion, it is possible to model the status and process in a quantitative model by integrating time-saving and cost-reducing methodological elements from remote sensing. The model developed in this study is based on general theories on vulnerability and substantiated with a set of theories. In terms of household vulnerability analysis, the majority of the theoretical set is derived from poverty and/or sustainability theories mainly aimed at the situations in developing or emerging countries. The theory set used for the evaluation of company vulnerability has its origin in a variety of empirical studies, mainly from developed countries. This set has been tested and developed further on the study regions in southern Thailand, representing a province with a lower middle income (Phang-Nga Province) and one with an upper middle income (Phuket) according to the WORLD BANK classification (WORLD BANK 2011). While the method of combining remote sensing data with household and company surveys in one structural equation model is generically applicable in general, the individual elements determining the perspectives of vulnerability need to be adapted carefully to the specific settings of the economic system there. Besides the setting of the economic system, the type of hazard also plays an important role, since this model has been developed for rapidly occurring covariant hazards such as tsunamis. With minor changes, therefore, it can be applied to storm floods or earthquakes and is thus suitable for multi-hazard analysis. The method is not applicable in this form to slowly occurring hazards, such as droughts, or idiosyncratic shocks, such as unemployment or the death of family members, where the specific location of the economic element does not play a crucial role and indirect effects deriving from outside influences are not a factor. Furthermore, it can be assumed that among the total population of the households and companies in the study regions, different interrelations within the vulnerability model exist. Therefore, a further step would be an extension of the database to use advanced methods such as the finite mixture approach that separates groups with differing interrelations from a sample. Due to the limited sample size, this evaluation method could not be applied to the dataset used in this study. An open question that could not be answered within this study is the extension of sensitivity and resilience elements to the totality of households and companies in the study region using remote sensing data. For this study, a calculation based on mean values on the community level was used. In future studies, it would be crucial to identify suitable methods in order to link further socio-economic characteristics to automatically derived households and companies using artificial neural networks or fuzzy logic methods. These self-learning networks can be trained to reproduce induced patterns with a given dataset and extend this pattern to a total population.

In addition to the impact on livelihoods and business activities, events such as natural disasters are always a starting point for an adaptation process which can change the society, its risk management and also its characteristics. This adaptation process can have various initiators, starting from individuals and companies, continuing up to non-governmental or governmental agencies. The development of adaptation strategies is subject to the experience from the natural disasters, but also, as WILLIAMSON (2000) argues, formal and informal institutions forming constraints that are given by the society. These institutions are not static either, but are subject to constant changes over time, which might be also triggered by the impact of natural hazards. Together with governance processes among the players, this dynamic system develops a variety of adaptation strategies which might also differ in their acceptance and effectiveness for the targeted groups. This study has taken a static picture of the adaptation strategies developed by the different players after the tsunami of 2004 and has attempted to evaluate their acceptance and effectiveness among seven different occupational groups, using qualitative focus group designs in order to identify the strategies and their outcome. In order to integrate these findings into a generic risk information tool, further steps are necessary, including the development of a dynamic risk governance framework integrating the elements mentioned above and the dynamic evaluation of the outcomes of a quantitative vulnerability model, as has been developed in this study. Therefore, a method has to be developed for measuring and evaluating adaptation strategies in a quantitative way and applying their outcomes to households and companies throughout the area.

References

ADGER W. N.; BROOKS N.; KELLY M.; BENTHAM, S.; ERIKSEN, S. (2004): *New indicators of vulnerability and adaptive capacity.* Tyndall Centre Technical Report 7. Norwich.

ADGER, W. N. (2006): *Vulnerability. Resilience, Vulnerability, and Adaptation: A Cross-Cutting Theme of the International Human Dimensions Programme on Global Environmental Change.* In: Global Environmental Change. 16(3), 268–281.

ADPC (Asian Disaster Preparedness Center) (2006): *The Economic Impact of the 26 December 2004 Earthquake & Indian Ocean Tsunami in Thailand.* Pathumthani.

ALDERMAN, H.; PAXSON, C. H. (1992): *Do the poor insure? A synthesis of the literature on risk and consumption in developing countries.* World Bank Agricultural policies 1008. Washington.

ALWANG, J.; SIEGEL, P. B.; JØRGENSEN, S. (2001): *Vulnerability: A View from Different Disciplines.* World Bank - Social Protection Discussion Paper 115. Washington D.C.

ALWANG, J.; SIEGEL, P. B.; JØRGENSEN, S. L. (1996): *Seeking guidelines for poverty reduction in rural Zambia.* In: World development. 24(11), 1711–1723.

AMANTE, C.; EAKINS, B. (2009): *ETOPO1 1 Arc-Minute Global Relief Model: Procedures, Data Sources and Analysis. NOAA Technical Memorandum NESDIS NGDC.* Washington D.C.

ARBUCKLE, J. L. (2006): *Amos 7.0 user's guide.* Chicago.

ATILGAN, E.; AKINCI, S.; AKSOY, S. (2003): *Mapping service quality in the tourism industry.* In: Managing Service Quality. 13(5), 412–422.

ATKINS, J. P.; MAZZI, S.; EASTER, C. D. (2000): *A Commonwealth vulnerability index for developing countries. The position of small states.* Commonwealth Secretariat - Economic paper 40. London.

BACKHAUS, K. (2008): *Multivariate Analysemethoden. Eine anwendungsorientierte Einführung.* Berlin.

BARRETT, C.; REARDON, T.; WEEB, P. (2001): *Nonfarm income diversification and household livelihood strategies in rural Africa: concepts, dynamics, and policy implications.* In: Food policy. 26(4), 315–331.

BARRON, D. N.; WEST, E.; HANNAN, M. T. (1994): *A Time to Grow and a Time to Die: Growth and Mortality of Credit Unions in New York City, 1914-1990.* In: The American Journal of Sociology. 100(2), 381–421.

BECK, U. (2009): *Risk society. Towards a new modernity.* London.

BENSON, C.; TWIGG, J.; MYERS, M. (2001): *NGO Initiatives in Risk Reduction: An Overview.* In: Disasters. 25(3), 199–215.

BEREKOVEN, L.; ECKERT, W.; ELLENRIEDER, P. (2009): *Marktforschung. Methodische Grundlagen und praktische Anwendung.* Wiesbaden.

BERTRAND, J. T.; BROWN, J. E.; WARD, V. M. (1992): *Techniques for Analyzing Focus Group Data.* In: Evaluation Review. 16(2), 198–209.

BEYENE, M. (1992): *Ein Informationssystem für die Abschätzung von Hochwasserschadenspotentialen.* (=Mitteilungen des Instituts für Wasserbau und Wasserwirtschaft der RWTH Aachen, 83). Aachen.

BILLING, P.; MADENGRUBER, U. (2006): *Overcoming the black hole: Outline for a quantitative model to compare coping capacities across countries.* In: BIRKMANN, J. (ed.): *Measuring vulnerability to natural hazards. Towards disaster resilient societies.* Tokyo, 403–415.

BIRD, K.; MCKAY, A.; SHINYEKWA, I. (2007): *Isolation and poverty: the relationship between spatially differentiated access to goods and services and poverty. Understanding and addressing spatial poverty traps.* Chronic Poverty Research Centre - Stellenbosch.

BIRKLAND, T. A.; HERABAT, P.; LITTLE, R. G.; WALLACE, W. A. (2006): *The Impact of the December 2004 Indian Ocean Tsunami on Tourism in Thailand.* In: Earthquake Spectra. 22(3), 889–900.

BIRKMANN, J. (2006): *Measuring vulnerability to promote disaster-resilient societies: Conceptual frameworks and definitions.* In: BIRKMANN, J. (ed.): *Measuring vulnerability to natural hazards. Towards disaster resilient societies.* Tokyo. 9–55

BIRKMANN, J. (2011): *First- and second-order adaptation to natural hazards and extreme events in the context of climate change.* In: Natural Hazards. doi: 10.1007/s11069-011-9806-8

BIRKMANN, J.; FERNANDO, N. (2008): *Measuring revealed and emergent vulnerabilities of coastal communities to tsunami in Sri Lanka.* In: Disasters. 32(1), 82–105.

BIRKMANN, J.; FERNANDO, N.; HETTIGE, S.; AMARASINGHE, S.; JAYASINGAM, T.; PARANAGAMA, D.; NANDANA, M.; NASSEL, M.; VOIGT, S.; GROTE, S.; ENGEL, S.; SCHRAVEN, B.; WOLFERTS, J. (2006): *Rapid and Multidimensional Vulnerability Assessment Sri Lanka at the Local Level.* Bonn.

BLAIKIE, P.; CANNON, T.; DAVIS, I.; WISNER, B. (2007): *At risk. Natural hazards, people's vulnerability and disasters.* London.

BLOOR, M. (2001): *Focus groups in social research.* London.

BOGARDI, J.; BIRKMANN, J. (2004): *Vulnerability Assessment. The First Steps towards Sustainable Risk Reduction.* In: MALZAHN, D.; PLAPP, T. (Ed.): *Disasters and society, from hazard assessment to risk reduction. Proceedings of the international conference.* Universität Karlsruhe (TH), Germany, July 26-27, 2004. Berlin. 75–82.

BOLLEN, K. A. (1989): *Structural equations with latent variables.* New York.

BOT (BANK OF THAILAND) (2010): *Labor Force Status of the Southern Region.* Bangkok.

BRIGUGLIO, L. (2003): *The Vulnerability Index And Small Island Developing States. A Review Of Conceptual and Methodological Issues.* AIMS Regional Preparatory Meeting. Praia.

BRISLIN, R. W. (2000): *Understanding culture's influence on behavior.* Fort Worth.

BROOKS, N. (2003): *Vulnerability, risk and adaptation: A conceptual framework.* Tyndall Centre Working Paper 38. Norwich.

BROSIUS, F. (2008): *SPSS 16.* Heidelberg.

BURNARD, P.; NAIYAPATANA, W.; LLOYD, G. (2006): *Views of mental illness and mental health care in Thailand: a report of an ethnographic study.* In: Journal of Psychiatric and Mental Health Nursing. 13(6), 742–749.

BURNS, W. (1993): *Covariance Structure Modeling: What We Have Learned From Simulation Studies and a Review of How Covariance Structure Models Have Been Treated In the Marketing Literature.* Marketing Science Conference. St. Louis.

BURNS, W.; SLOVIC, P.; KASPERSON, R.; KASPERSON, J.; RENN, O.; EMANI, S. (1993): *Incorporating Structural Models into Research on the Social Amplification of Risk: Implications for Theory Construction and Decision Making.* In: Risk Analysis. 13 (6), 611–623.

BURTON, I.; KATES, R. W.; WHITE, G. F. (1993): *The environment as hazard.* New York.

BWK (Bund der Ingenieure für Wasserwirtschaft, Abfallwirtschaft und Kulturbau) (2001): *Hochwasserschadenspotentiale.* Bericht 1/2001. Düsseldorf.

BYRNE, B. M. (2001): *Structural equation modeling with AMOS. Basic concepts, applications, and programming.* Mahwah.

CALGARO, E. (2005): *Paradise in Tatters: An Analysis of the Vulnerability of the Tourism Community of Khao Lak, Thailand, to the 2004 Indian Ocean Tsunami.* Stockholm.

CALGARO, E.; LLOYD, K. (2008): *Sun, sea, sand and tsunami: examining disaster vulnerability in the tourism community of Khao Lak, Thailand.* In: Singapore Journal of Tropical Geography. 29(3), 288–306.

CALGARO, E.; NARUCHAIKUSOL, S.; PONGPONRAT, K. (2009): *Comparative Destination Vulnerability Assessment for Khao Lak, Patong Beach and Phi Phi Don. Sustainable Recovery and Resilience Building in the Tsunami Affected Region.* Stockholm Environment Institute. Stockholm.

CAMERON, J. (2005): *Focusing on the focus group.* In: HAY, I. (ed.): *Qualitative research methods in human geography.* New York. 156–174.

CANNON, T.; TWIGG, J.; ROWELL, J. (2003): *Social Vulnerability. Sustainable Livelihoods and Disasters.* Report to DFID Conflict and Humanitarian Assistance Department and Sustainable Livelihoods Support Office. London.

CARDONA, O. D. (1999): *Environmental Management and Disaster Prevention. Two Related Topics: A Holistic Risk Assessment and Management Approach.* In: INGLETON, J. (Hg.*): Natural disaster management. A presentation to commemorate the International Decade for Natural Disaster Reduction (IDNDR), 1990-2000.* Leicester. 686–695.

CARDONA, O. D. (2001): *Estimación Holística del Riesgo Sísmico utilizando Sistemas Dinámicos Complejos.* Universidad Politécnica de Cataluña. Barcelona. Online available: http://www.desenredando.org/public/varios/2001/ehrisusd/index.html.

CARLSEN, J. C.; HUGHES, M. (2008): *Tourism Market Recovery in the Maldives after the 2004 Indian Ocean Tsunami.* In: Journal of Travel & Tourism Marketing. 23(2), 139–149.

CARTER, M. R.; LITTLE, P. D.; MOGUES, T.; NEGATU, W. (2004): *Shocks, Sensitivity and Resilience: Tracking the Economic Impacts of Environmental Disaster on Assets in Ethiopia and Honduras.* Basis - Colloborative Research Program. Madison.

CASSEL, C. (1999): *Robustness of partial least-squares method for estimating latent variable quality structures.* In: Journal of applied statistics. 26(4), 435–446.

CHAIPAN, C.; CHAIPAN, C. (2006): *The shy birds are making a comeback, but... An economic and cultural evaluation of Thailand's tourism industry after the tsunami.* In: Pacific News. (25), 4–7.

CHAMBERS, R. (1989): *Editorial Introduction: Vulnerability, Coping and Policy.* In: IDS Bulletin. 20(2), 1–7.

CHANG, S. (2000): *Disasters and transport systems: loss, recovery and competition at the Port of Kobe after the 1995 earthquake.* In: Journal of Transport Geography 8(1), 53–65.

CHATENOUX, B.; PEDUZZI, P. (2007): *Impacts from the 2004 Indian Ocean Tsunami: analysing the potential protecting role of environmental features.* In: Natural Hazards. 40(2), 289–304.

CHAUDHURI, S. (2003): *Assessing vulnerability to poverty: concepts, empirical methods and illustrative examples.* Washington D.C.

CHAUDHURI, S.; JALAN, J.; SURYAHADI, A. (2002): *Assessing household vulnerability to poverty from cross-sectional data a methodology and estimates from Indonesia.* Department of Economics, Columbia University - Discussion paper 0102-52. New York.

CHIN, W. W. (2009): *The Partial Least Squares Approach for Structural Equation Modeling.* In: MARCOULIDES, G. A. (ed.): *Modern methods for business research. Quantitative methodology series.* New York. 295–336.

CHIN, W. W.; NEWSTED, P. R. (1999): *Structural Equation Modeling Analysis with small Samples using Partial Least Squares.* In: HOYLE, R. H. (ed.): *Statistical strategies for small sample research.* Thousand Oaks. 307–342.

CLARK, W.; JAEGER, J.; CORELL, R.; KASPERSON, R.; MCCARTHY, J.; CASH, D.; COHEN, S.; DESANKER, P.; DICKSON, N.; EPSTEIN, P.; GUSTON, D.; HALL, J.; JAEGER, C.; JANETOS, A.; LEARY, N.; LEVY, M. (2000): *Assessing Vulnerability to Global Environmental Risks.* Belfer Centre - Discussion paper 12. Cambridge.

CLARKE, G. R. G.; WALLSTEN, S. J. (2003): *Do Remittances Act Like Insurance? Evidence From a Natural Disaster in Jamaica.* In: SSRN Electronic Journal. doi:10.2139/ssrn.373480

CLINTON, W. J. (2006): *Lessons Learned from Tsunami Recovery. Key Propositions for Building Back Better. United Nations Secretary-General's Special Envoy for Tsunami Recovery.* New York.

COCHRANE, H. (2003): *Economic Loss. Myth and Measurement.* In: VAN DER VEEN, A.; VETERE ARELLANO, A.; NORDVIK, J. (eds.): *In search of a common methodology on damage estimation.* Twente, 9–18.

CONZE, O. (2007): *Kundenloyalität durch Kundenvorteile. Segmentspezifische Analyse und Implikationen für das Kundenbeziehungsmanagement.* Wiesbaden.

COUDOUEL, A.; HENTSCHEL, J. S.; WODON, Q. T. (2002): *Poverty Measurement and Analysis.* In: WORLD BANK (ed.): *Poverty Reduction Strategy Sourcebook.* Washington D.C., 29–74.

CUTTER, S. L.; BORUFF, B. J.; SHIRLEY, W. L. (2003): *Social Vulnerability to Environmental Hazards.* In: Social Science Quarterly. 84(2), 242–261.

DACY, D. C.; KUNREUTHER, H.; WILLIAMSON, H. A. (1969): *The economics of natural disasters. Implications for federal policy.* New York.

DAHLHAMER, J.; TIERNEY, K. (1998): *Rebounding from disruptive events: Business recovery following the Northridge earthquake.* In: Sociological Spectrum. 18(2), 121–141.

DBP (Department of Business Development) (2011): *Companies in Thailand.* Bangkok

DDPM (Department for Disaster Preparedness and Mitigation) (2005a): *Earthquake/Tsunami Victims Relief Efforts.* Bangkok.

DDPM (Department for Disaster Preparedness and Mitigation) (2005b): *Loss Assessment by Provinces from Tsunami and Earthquake.* Bangkok.

DDPM (Department for Disaster Preparedness and Mitigation) (2007): *Risk areas from the tsunami in the area six Andaman coastal provinces. Phang-Nga Province.* Bangkok.

DEJANVRY, A.; FAFCHAMPS, M.; SADOULET, E. (1991): *Peasant household behaviour with missing markets. Some paradoxes explained.* In: The Economic Journal. 1991, 1400–1417.

DESHINGKAR, P.; AHEEYAR, M. (2006): *Remittances in crisis. Sri Lanka after the Tsunami.* HPG Background Paper. London.

DIAMANTOPOULOS, A.; WINKLHOFER, H. M. (2001): *Index Construction with Formative Indicators: An Alternative to Scale Development.* In: Journal of Marketing Research. 38(2), 269–277.

DIEKMANN, A. (2010): *Empirische Sozialforschung. Grundlagen, Methoden, Anwendungen.* Reinbek

DIENER, E.; SUH, E. M. (2000): *Culture and subjective well-being.* Cambridge.

DILLEY, M. (2005): *Natural disaster hotspots. A global risk analysis.* Washington (D.C.).

DIXON, C. J. (1999): *The Thai economy. Uneven development and internationalization.* New York.

DOUGLAS, J. D. (1976): *Investigative social research. Individual and team field research.* London.

DUYNE BARENSTEIN, J. (2006): *Housing reconstruction in post-earthquake Gujarat. A comparative analysis.* Humanitarian Practice Network Paper 54. London.

DWYER, L. (2000): *The price competitiveness of travel and tourism: a comparison of 19 destinations.* In: Tourism Management. 21(1), 9–22.

EAKIN, H. (2005): *Institutional change, climate risk, and rural vulnerability: Cases from Central Mexico.* In: World Development 33(11), 1923–1938.

ECLAC (Economic Commission for Latin America and the Caribbean) (2003): *Handbook for Estimating the Socio-economic and Environmental Effects of Disasters.* Santiago de Chile

EICKELPASCH, A.; LEIPRAS, A.; STEPHAN, A. (2007): *Hard and Soft Locational Factors, Innovativeness and Firm Performance: An Empirical Test of Porter's Diamond Model at the Micro-Level.* DIW Discussion Papers 109. Berlin

EM-DAT (2010): *The OFDA/CRED International Disaster Database.* http://www.emdat.be 10.04.2010.

ESPON (European Spatial Planning Observation Network) (2003): Glossary of Terms. http://www.gsf.fi/projects/espon/glossary.htm. 14.06.2010

FALVEY, L. (2000): *Thai agriculture. Golden cradle of millennia.* Bangkok.

FAO (Food and Agriculture Organization) (2005a): *Report of Joint FAO/MOAC Detailed Technical Damages and Needs Assessment Mission in Fisheries and Agriculture Sectors in Tsunami Affected Six Provinces in Thailand.* Rome.

FAO (Food and Agriculture Organization) (2005b): *Salt-Affected Soils from sea water intrusion. Strategies for Rehabilitation and Management.* Rome.

FAO (Food and Agriculture Organization) (2006): *Fact Sheet: The international fish trade and world fisheries.* Rome.

FASSOT, G.; EGGERT, A. (2005): *Zur Verwendung formativer und reflektiver Indikatoren in Strukturgleichungsmodellen. Bestandsaufnahme und Anwendungsempfehlungen.* In: BLIEMEL, F. (ed.): *Handbuch PLS-Pfadmodellierung : Methode, Anwendung, Praxisbeispiele.* Stuttgart. 31-49

FLAX, L. K.; JACKSON, R. W.; STEIN, D. N. (2002): *Community Vulnerability Assessment Tool Methodology.* In: Natural Hazards Review 3(4), 163–176.

FLINT, M.; GOYDER, H. (2006): *Funding the tsunami response.* London.

FOLKE, C. (2006): *Resilience: The emergence of a perspective for social–ecological systems analyses.* In: Global Environmental Change. 16(3), 253–267.

FORNELL, C.; BOOKSTEIN, F. (1982): *A Comparative analysis of two structural equation models : LISREL and PLS applied to market data.* In: FORNELL, C. (ed.): *A second generation of multivariate analysis.* New York, 289–324.

FREYER, W. (2009): *Tourismus. Einführung in die Fremdenverkehrsökonomie. Lehr- und Handbücher zu Tourismus, Verkehr und Freizeit.* München.

GLASSMAN, J.; SNEDDON, C. (2003): *Chiang Mai and Khon Kaen as Growth Poles: Regional Industrial Development in Thailand and its Implications for Urban Sustainability.* In: The Annals of the American Academy of Political and Social Science. 590(1), 93–115.

GOODWIN, H. (2007): *Measuring and Reporting the Impact of Tourism on Poverty.* In: TRIBE, J.; AIREY, D. (eds.): *Developments in tourism research.* Amsterdam. 63–75.

GROVES, R. M. (2009): *Survey methodology.* Hoboken.

GUHA SAPIR, D.; JOSHI, P. C.; ARNOLD, J. P. S.; DEGOMME, O.; PARRY, L. (2006): *Risk factors for mortality and injury. Post-tsunami epidemiological findings from Tamil Nadu.* Brussels.

GUIMARÃES, R. J. (2007): *Searching for the Vulnerable: A Review of the Concepts and Assessments of Vulnerability Related to Poverty.* In: The European Journal of Development Research. 19(2), 234–250.

GURTNER, Y.K (2007): *Phuket: Tsunami and Tourism - a Preliminary Investigation.* IN: LAWS, E.; PRIDEAUX B.; CHON, K. S (eds.): *Crisis management in tourism.* Wallingford. 217–233.

HÄDER, M. (2010): *Empirische Sozialforschung. Eine Einführung.* Wiesbaden.

HAGIWARA, A. T.; SUGIYARTO, G. (2005): *Poverty Impact of the Tsunami: An initial assessment and scenario analysis.* Paper presented during the 4th PEP Research Network General Meeting. Colombo.

HARBITZ, C.; SVERDRUP-THYGESON, K.; KAISER, G.; SWARNY, R.; GRUENBURG, L.; GLIMSDAL, S.; LOVHOLT, F.; MCADOO, B.; FRAUENFELDER, R. (2011): *GIS methodologies for local tsunami risk assessment – validation for the 2009 South Pacific tsunami in American Samoa.* EGU General Assembly 2011. Vienna.

HEINZ CENTER (2000): *The hidden costs of coastal hazards: implications for risk assessment and mitigation.* Washington D.C.

HEITZMANN, K.; CANAGARAJAH, R. S.; SIEGEL, P. B. (2002): *Guidelines for Assessing the Sources of Risk and Vulnerability.* World Bank - Social Protection Discussion Paper 218. Washington D.C.

HERRMANN, A.; HUBER, F.; KRESSMANN, F. (2006): *Varianz- und kovarianzbasierte Strukturgleichungsmodelle - ein Leitfaden zu deren Spezifikation, Schätzung und Beurteilung.* In: Zeitschrift für betriebswirtschaftliche Forschung. 58(2), 34–66.

HOLZMANN, R.; JØRGENSEN, S. (2001): *Social Risk Management: A New Conceptual Framework for Social Protection, and Beyond.* In: International Tax and Public Finance. 8(4), 529–556.

HOOGEVEEN, H. (2000). *Risk and insurance by the poor in developing countries.* Paper presented at the "Microfinance for Disaster Risk Colloquium." WORLD BANK. WASHINGTON, D.C.

HORWICH, G. (2000): *Economic Lessons of the Kobe Earthquake.* In: Economic Development and Cultural Change. 48(3), 521–542.

HUANG, J.-H.; CHUANG, S.-T.; LIN, Y.-R. (2008): *Folk Religion and Tourist Intention. Avoiding Tsunami-Affected Destinations.* In: Annals of Tourism Research. 35(4), 1074–1078.

HUBER, F.; HERRMANN, A.; MEYER, F.; VOGEL, J.; VOLLHARDT, K. (2007): *Kausalmodellierung mit Partial Least Squares. Eine anwendungsorientierte Einführung.* Wiesbaden.

ICHINOSAWA, J. (2006): *Reputational disaster in Phuket: the secondary impact of the tsunami on inbound tourism.* In: Disaster Prevention and Management. 15(1), 111–123.

ILO (International Labour Organization) (2005): *After the tsunami In Thailand, the tourist industry fights back.* Geneva

IMF (International Monetary Fund) (2006): Thailand*: Report on the Observance of Standards and Codes—Data Module, Response by the Authorities, Detailed Assessments Using the Data Quality Assessment Framework (DQAF).* IMF Country Report 06/140. Washington D.C.

INTUWONGES, S. (2010) *Non-Life Sector in Brief. Thai Reinsurance.* Bangkok

IOUALALEN, M.; ASAVANANT, J.; KAEWBANJAK, N.; GRILLI, S. T.; KIRBY, J. T.; WATTS, P. (2007): *Modeling the 26 December 2004 Indian Ocean tsunami: Case study of impact in Thailand.* In: Journal of Geophysical Research. 112(C7), C07024

IPCC (Intergovernmental Panel on Climate Change) (2001): *Climate Change 2001: Synthesis Report.* Geneva.

IPCC (Intergovernmental Panel on Climate Change) (2007): *Climate Change 2007: Synthesis Report. Contribution of Working Groups I, II and III to the Fourth Assessment Report of the Intergovernmental Panel on Climate Change.* Geneva.

ISARD, W. (1998): *Methods of interregional and regional analysis.* (=Regional science studies series). Aldershot.

JANKAEW, K.; ATWATER, B. F.; SAWAI, Y.; CHOOWONG, M.; CHAROENTITIRAT, T.; MARTIN, M. E.; PRENDERGAST, A. (2008): *Medieval forewarning of the 2004 Indian Ocean tsunami in Thailand.* In: Nature. 455(7217), 1228–1231.

JARVIS, C. B.; MACKENZIE, S. B.; PODSAKOFF, P. M. (2003): *A Critical Review of Construct Indicators and Measurement Model Misspecification in Marketing and Consumer Research.* In: Journal of Consumer Research. 30(2), 199–218.

JAYASURIYA, S.; MCCAWLEY, P. (2008): *Reconstruction after a Major Disaster. Lessons from the Post-Tsunami Experience in Indonesia, Sri Lanka, and Thailand.* ADBI Working Paper 125. Tokyo.

KAMENZ, U. (2001): *Marktforschung. Einführung mit Fallbeispielen, Aufgaben und Lösungen.* Stuttgart.

KAMOLVEJ, T. (2006): *The integration of intergovernmental coordination and information management in response to immediate crises. Thailand emergency management.* Dissertation. Pittsburgh.

KASPERSON, J.; KASPERSON, R.; TURNER, B. L.; HSIEH, W.; SCHILLER, A. (2005): *Vulnerability to Global Environmental Change.* In: KASPERSON, R.; KASPERSON, J. (eds.): *Social Contours of Risk, 2. Risk Analysis Corporations and the Globalization of Risk.* London, 245–285.

KASPERSON, R.; JHAVERI, N.; KASPERSON, J. X. (2004): *Stigma and the social amplification of risk: towards a framework of analysis.* In: FLYNN, J. (ed.): *Risk, media, and stigma. Understanding public challenges to modern science and technology.* London, 9–27.

KATHIRESAN, K.; RAJENDRAN, N. (2005): *Coastal mangrove forests mitigated tsunami.* In: Estuarine, Coastal and Shelf Science. 65(3), 601–606.

KEPPER, G. (2008): *Methoden der qualitativen Marktforschung.* In: HERRMANN, A. (ed.): *Handbuch Marktforschung. Methoden, Anwendungen, Praxisbeispiele.* Wiesbaden, 159–202.

KINSEY, B. (1998): *Coping with drought in Zimbabwe: Survey evidence on responses of rural households to risk.* In: World Development. 26(1), 89–110.

KLINE, R. B. (2005): *Principles and practice of structural equation modeling. Methodology in the social sciences.* New York, NY.

KNOX, P.L.; MARSTON, S.A (2008): *Humangeographie.* Heidelberg.

KORFF, R. (2009): *Der Weihnachts-Tsunami in Khao Lak. Eine abgeschlossene Katastrophe.* In: HAMMERL, C. (ed.): *Naturkatastrophen. Rezeption - Bewältigung - Verarbeitung.* (=Konzepte und Kontroversen 7). Innsbruck, 124–137.

KRAAS, F. (1996): *Bangkok. Ungeplante Megastadtentwicklung durch Wirtschaftsboom und soziokulturelle Persistenzen.* In: Geographische Rundschau. 48(2), 89–96.

KRACK, R. (2009): *Phuket und Umgebung.* Bielefeld.

KROLL, C. A., LANDIS, J. D., SHEN, Q., AND STRYKER, S. (1991): *Economic impacts of the Loma Prieta earthquake: A focus on small business.* University of California, Transportation Center and the Center for Real Estate and Economics - Working Paper No. 91-187. Berkeley.

LARSEN, R. K.; CALGARO, E.; THOMALLA, F. (2011): *Governing resilience building in Thailand's tourism-dependent coastal communities: Conceptualising stakeholder agency in social–ecological systems.* In: Global Environmental Change. doi: 10.1016/j.gloenvcha.2010.12.009

LIGNON, E.; SCHECHTER, L. (2003): *Measuring Vulnerability.* In: The Economic Journal. 113, 95–102.

LIOU, F.-M.; DING, C.G. (2004): *Positioning the non-least-developed developing countries based on vulnerability-related indicators.* In: Journal of International Development. 16(6), 751–767.

LOHMÖLLER, J.-B. (1989): *Latent variable path modeling with partial least squares.* Heidelberg.

LÜBBEN, I. (2009): *Migration und Rücküberweisungen in Thailand.* Dissertation. Hannover.

MARSH, H. W.; WEN, Z.; HAU, K. (2006): *Structural Equation Models of Latent Interation and Quadratic Effects.* In: HANCOCK, G. R.; MUELLER, R.O. (eds.): *Structural equation modeling. A second course. Quantitative methods in education and the behavioral sciences.* Greenwich. 225–269

MASLOW, A. H. (1970): *A theory of human motivation.* In: SEXTON, W. P. (ed.): *Organization theories.* Columbus, Ohio. 143–166.

MASSMANN, F. (2010): *Analyse der Vulnerabilität von Landwirtschaft und Fischerei an der Andamanküste Thailands im Kontext des Tsunamis von 2004.* Diploma Thesis. Kiel.

MAUSCH, K. (2010): *Poverty, inequality and the non-farm economy.* Dissertation. Hannover.

MAYRING, P. (2002): *Einführung in die qualitative Sozialforschung. Eine Anleitung zu qualitativem Denken.* Weinheim.

MCDOWALL, S.; WANG, Y. (2009): *An Analysis of International Tourism Development in Thailand: 1994-2007.* In: Asia Pacific Journal of Tourism Research. 14(4), 351–370.

MECHLER, R. (2004): *Natural disaster risk management and financing disaster losses in developing countries.* (=Karlsruher Reihe Risikoforschung und Versicherungsmanagement 2;1). Karlsruhe.

MEYER, V. (2005): *Methoden der Sturmflut-Schadenspotenzialanalyse an der deutschen Nordseeküste.* Dissertation. Leipzig.

MEYER, V.; MAI, S. (2003): *Verfahren zur Berechnung der Schäden nach Deichbruch an der deutschen Nordseeküste.* In: KELLETAT, D. (ed.): *Neue Ergebnisse der Küsten- und Meeresforschung. Tagungsband der 21. Jahrestagung des Arbeitskreises "Geographie der Meere und Küsten" (AMK) in Essen.* (=Essener Geographische Arbeiten 35). Essen.

MIDDLETON, V. T. C.; FYALL, A.; MORGAN, M. (2009): *Marketing in travel and tourism.* Amsterdam.

MILETI, D. S. (1999): *Disasters by design. A reassessment of natural hazards in the United States.* Washington.

MOI (Ministry of Interior) (2003): *SME Definition.* Bangkok

MONTZ, B. (1992): *The Effects of Flooding on Residential Property Values in Three New Zealand Communities.* In: Disasters. 16(4), 283–298.

MORDUCH, J. (1999): *Between the State and the Market: Can Informal Insurance Patch the Safety Net?* In: World Bank Research Observer. 14(2), 187–207.

MOSER, C. (1998): The asset vulnerability framework: Reassessing urban poverty reduction strategies. Reassessing urban poverty reduction strategies. IN: World Development 26(1), 1–19.

MPND (Ministry of Planning and National Development) (2009): *Statistical Yearbook 2009. Gross Domestic Product, by kind of activity, 2003-2009.* Malé.

MUNICHRE (2002): *MunichRe Group Annual Report 2002.* München

MYDANS, S. (2005): *After Tsunami, rebuilding paradise.* In: The New York Times. 24.04.2005

MYERS, G. (1998): *Displaying Opinions: Topics and Disagreement in Focus Groups.* In: Language in Society. 27(1), 85–111.

NACA (Network of Aquaculture Centres in Asia-Pacific) (2005): *Tsunami impact on fisheries and aquaculture in Thailand.* Bangkok.

NAGY, M. S. (2002): *Using a single-item approach to measure facet job satisfaction.* In: Journal of Occupational and Organizational Psychology. 75(1), 77–86.

NAUDÉ, W. A.; SANTOS-PAULINO, A. U.; MCGILLIVRAY, M. (2009): *Vulnerability in developing countries.* Tokyo.

NAUDÉ, W.; MCGILLIVRAY, M.; ROSSOUW, S. (2007): *A Local Vulnerability Index for South Africa. Conference Paper: 1st Annual Conference: Charlotte Manye Maxeke Collaboration in Research and Capacity Building in Economics of Social Protection.* Pilanesberg.

NAUDÉ, W.; MCGILLIVRAY, M.; ROSSOUW, S. (2008): *Measuring the vulnerability of subnational regions.* UNU-WIDER Research paper 2008/54. Helsinki.

NESDB (National Economic and Social Development Board) (1987): *The Sixth National Economic and Social Development Plan (1987–91).* Bangkok.

NESDB (National Economic and Social Development Board) (1992): *The Seventh National Economic and Social Development Plan (1992–96).* Bangkok.

NESDB (National Economic and Social Development Board) (2006): *Input Output Table 2005.* Bangkok.

NESDB (National Economic and Social Development Board) (2009): *Gross Regional and Provincial Products 2008.* Bangkok.

NESDB (National Economic and Social Development Board) (2010): *National Income of Thailand.* Bangkok.

NIDHIPRABHA, B. (2007): *Adjustment and Recovery in Thailand Two Years after the Tsunami.* ADB Institute Discussion Paper 72. Tokyo.

NOAA (National Oceanic and Atmospheric Administration) (2007): *Physics of Tsunamis.* Washington D.C.

NOAA (National Oceanic and Atmospheric Administration) (2010): *NO-AA/WDC Historical Tsunami Database at NGDC.* Washington D.C.

NRC (National Research Council (U.S.– Committee on Assessing the Costs of Natural Disasters) (1999): *The impacts of natural disasters. A framework for loss estimation.* Washington.

NSO (National Statistical Office) (2000): *The 2000 Population and Housing Census.* Bangkok

NSO (National Statistical Office) (2003): *2003 Agricultural Census - Southern Region.* Bangkok.

NSO (National Statistical Office) (2004): *Provincial Statistical Report 2003 - Phuket.* Bangkok.

NSO (National Statistical Office) (2007a)*: Provincial Statistical Report 2007 - Phuket.* Bangkok.

NSO (National Statistical Office) (2007b): *Business Trade and Service Survey 2007.* Bangkok.

NSO (National Statistical Office) (2007c): *Provincial Statistical Report 2007 - Phang-Nga.* Bangkok.

NSO (National Statistical Office) (2007d): *Provincial Statistical Report: 2007 - Phuket.* Bangkok.

NSO (National Statistical Office) (2007e): *The 2007 Household Socio-Economic survey - Southern Region.* Bangkok.

NSO (National Statistical Office) (2008a): *Key Statistics Thailand 2007.* Bangkok.

NSO (National Statistical Office) (2008b): *The 2007 Informal Employed Survey.* Bangkok.

OEM (Office of Emergency Preparedness) (1972): *Disaster preparedness. Report to the Congress.* Washington.

OKUYAMA, Y. (2003): *Modeling Spatial Economic Impacts of an Earthquake: Input-Output Approaches.* In: VAN DER VEEN, A.; VETERE ARELLANO, A.; Nordvik J.-P. (eds.): In *search of a common methodology on damage estimation.* Twente, 113–137.

OORT, F. G. VAN; OUD, J. H. L.; RASPE, O. (2009): *The urban knowledge economy and employment growth: a spatial structural equation modeling approach.* In: The Annals of Regional Science. 43(4), 859–877.

OOSTERHAVEN, J.; STELDER, D. (2007): *Regional and Interregional IO Analysis.* Groningen.

PANJARAT, S. (2008): *Sustainable Fisheries in the Andaman Sea Coast of Thailand.* New York.

PANTEN, G.; BOßOW-THIES, S. (2007): *Analyse kausaler Wirkungszusammenhänge mit Hilfe von Partial Least Squares (PLS).* In: ALBERS, S.; KLAPPER, D.; KONRADT, U.; WALTER, A.; WOLF, J. (eds.): *Methodik der empirischen Forschung.* Wiesbaden, 311–326.

PARNWELL, M. (1992): *Confronting uneven development in Thailand: The potential role of rural industries.* In: Malaysian Journal of Tropical Geography. 22(1), 51–62.

PATON, D.; GREGG, C. E.; HOUGHTON, B. F.; LACHMAN, R.; LACHMAN, J.; JOHNSTON, D. M.; WONGBUSARAKUM, S. (2008): *The impact of the 2004 tsunami on coastal Thai communities: assessing adaptive capacity.* In: Disasters. 32(1), 106–119.

PELLING, M.; UITTO, J. I. (2001): *Small island developing states: natural disaster vulnerability and global change.* In: Global Environmental Change Part B: Environmental Hazards. 3(2), 49–62.

PETCHA-EM, M. (2008): *Thailand Land Administration. International Seminar on Land Administration Trends and Issues in the Asia and Pacific Region.* Kuala Lumpur.

PHAPHAVASIT, N.; AKSORNKOAE, S.; SILVA, J. DE (2009): *Tsunami Impact on Mangrove Ecosystems.* Nonthaburi.

RAMCHARAN, R. (2005): *How big are the benefits of economic diversification? Evidence from earthquakes.* IMF working paper 05/48. Washington, DC.

RATTANANAKIN, S. (2006): *Summary report of assistance given to Tsunami affected victims.* Bangkok.

RAYHAN, I. (2008): *Assessing household vulnerability and coping strategies to floods.* Dissertation. Hannover.

REARDON, T.; MATLON, P.; DELGADO, C. (1988): *Coping with household-level food insecurity in drought-affected areas of Burkina Faso.* In: World development. 16(9), 1065–1074.

REESE, S. (2003): *Die Vulnerabilität des schleswig-holsteinischen Küstenraumes durch Sturmfluten - Fallstudien von der Nord- und Ostseeküste.* (=Berichte aus dem Forschungs- und Technologiezentrum Westküste der Universität Kiel 30). Kiel

REESE, S.; COUSINS, W. J.; POWER, W. L.; PALMER, N. G.; TEJAKUSUMA, I. G.; NUGRAHADI, S. (2007): *Tsunami vulnerability of buildings and people in South Java; field observations after the July 2006 Java tsunami.* In: Natural Hazards and Earth System Science. 7(5), 573–589.

RÉGNIER, P.; NERI, B.; SCUTERI, S.; MINIATI, S. (2008): *From emergency relief to livelihood recovery: Lessons learned from post-tsunami experiences in Indonesia and India.* In: Disaster Prevention and Management. 17(3), 410–430.

RENN, O. (2008): *Risk governance. Coping with uncertainty in a complex world.* London.

REVENUE DEPARTMENT (2008): *Personal Income Tax.* Bangkok. http://www.rd.go.th/publish/6045.0.html 14.06.2010

REVILLA DIEZ, J. (1998): *Bedeutung und regionale Verteilung ausländischer Direktinvestitionen in der SR Vietnam.* In: Zeitschrift für Wirtschaftsgeographie. 42(3/4), 201–212.

RICHARDSON, H. (1980): *Polarization Reversal in Developing Countries.* In: Papers in Regional Science. 45(1), 67–85.

RITTICHAINUWAT, B. (2006): *Tsunami Recovery: A Case Study of Thailand's Tourism.* In: Cornell Hotel and Restaurant Administration Quarterly. 47(4), 390–404.

RITTICHAINUWAT, B. (2011): *Ghosts.* In: Annals of Tourism Research. 38(2), 437–459.

ROEMER, H.; KAISER, G.; STERR, H.; LUDWIG, R. (2010): *Using remote sensing to assess tsunami-induced impacts on coastal forest ecosystems at the Andaman Sea coast of Thailand.* In: Natural. Hazards and Earth Systems Science. 10(4), 729–745.

RoM (Republic of Maldives) (2005): *Tsunami: Impact and Recovery.* Joint Needs Assessment. Malé

ROSE, A.; GUHA, G. (2004): *Computable General Equilibrium Modeling of Electric Utility Lifeline Losses from Earthquakes.* In: OKUYAMA, Y. (ed.): *Modeling spatial and economic impacts of disasters. Advances in spatial science.*, 119–143.

RUANGRASSAMEE, A.; YANAGISAWA, H.; FOYTONG, P.; LUKKUNAPRASIT, P.; KOSHIMURA, S.; IMAMURA, F. (2006): *Investigation of Tsunami-Induced Damage and Fragility of Buildings in Thailand after the December 2004 Indian Ocean Tsunami.* In: Earthquake Spectra. 22(S3), 377-401.

SANTANA, G. (2004): *Crisis Management and Tourism.* In: Journal of Travel & Tourism Marketing 15(4), 299–321.

SCHÄTZL, L. (2000): *Empirie. 2.* Paderborn

SCHÄTZL, L. (2003): *Wirtschaftsgeographie.* Paderborn.

SCHILLER, D. (2006): *Universitäre Industriekooperationen in Thailand. Auswirkungen des Wandels im thailändischen Hochschulsystem auf Kommerzialisierungsstrategien der Hochschulen und Wissenstransfer im Innovationssystem.* (=Wirtschaftsgeographie 37). Berlin.

SCHNEIDERBAUER, S. (2008): *Risk and Vulnerability to Natural Disasters - from Broad View to Focused Perspective. Theoretical background and applied methods for the identification of the most endangered populations in two case studies at different scales.* Dissertation. Berlin.

SCHULTZ, T. W. (1975): *Human Capital and Disequilibrium.* In: Journal of Economic Literature. (13) 827–846.

SCHWARZE, S. (2004): *Determinants of Income Generating Activities of Rural Households. A Quantitative Study in the Vicinity of the Lore-Lindu National Park in Central Sulawesi/Indonesia.* Dissertation. Göttingen.

SCHWARZE, S.; ZELLER, M. (2005): *Income diversification of rural households in Central Sulawesi, Indonesia.* In: Quarterly journal of international agriculture. 44(1), 61–73.

SCOONES, I. (1998): *Sustainable Rural Livelihoods. A Framework for Analysis.* IDS Working Paper 72. Brighton.

SEGSCHNEIDER, K.; CHAIKAO, T.; MAREOH, R. (2008): *The Ecology and Environment of Bahn Nahm Khem Three Years after the Tsunami. Tsunami Aid Watch.* Chiang Mai.

SEN, A. (1981): *Poverty and famines. An essay on entitlement and deprivation.* Oxford.

SIEGEL, P. B.; ALWANG, J.; JØRGENSEN, S. L. (2003): *Rediscovering vulnerability through a risk chain. Views from different disciplines.* In: Quarterly journal of international agriculture. 42(3), 351–370.

SIEGEL, P.; ALWANG, J. (1999): *An Asset-Based Approach to Social Risk Management. A Conceptual Framework.* Social Protection Discussion Paper 9926. Washington D.C.

SIM, B.; SECRETARIO, F.; SUAN, E. (2007): *Developing an interregional input-output table for cross-border economics - An application to Lao People's Democratic Republic and Thailand.* ERD Occasional Statistical Paper Series No. 1. Manila.

SINGBUN, P.; OPASPAKORNKIT, S.; SIRISUK, D.; JONGKAIJAK, M.; SUKSAI, S.; TO-RANG, S. (2008): *Lessons learned from Thailand's tsunami experiences for future preparedness: 60 case studies of good practice for Thailand disaster management (2004–2007) from Tsunami Knowledge Decode Project.* Bangkok.

SMITH, K.; WARD, R. (1998): *Floods. Physical processes and human impacts.* Chichester.

SOLESBURY, W. (2003): *Sustainable livelihoods. A case study of the evolution of DFID policy.* DFID Working Paper 217. London.

SONMEZ, S. F.; APOSTOLOPOULOS, Y.; TARLOW, P. (1999): *Tourism in Crisis: Managing the Effects of Terrorism.* In: Journal of Travel Research. 38(1), 13–18.

STELDER, D.; OOSTERHAVEN, J.; EDING, G. J. (2000): *Interregional Input-Output Software. IRIOS 1.0 Manual.* Groningen.

STERR, H.; KAISER, G.; ROEMER, H.; REVILLA DIEZ, J.; WILLROTH, P.; LUDWIG, R. (2009): *Tsunamigefährdung an den Küsten Thailands : Was lehrt die Katastrophe von 2004?* In: Geographische Rundschau. 61(12), 28–35.

SUH, Y. (2004): *Preferences and trip expenditures—a conjoint analysis of visitors to Seoul, Korea.* In: Tourism Management. 25(1), 127–137.

TAMÁSY, C. (2005): *Determinanten des regionalen Gründungsgeschehens.* (=Wirtschaftsgeo-graphie 27). Münster.

TAN-MULLINS, M.; RIGG, J.; LAW, L.; GRUNDY-WARR, C. (2007): *Re-mapping the politics of aid: the changing structures and networks of humanitarian assistance in post-tsunami Thailand.* In: Progress in Development Studies. 7(4), 327–344.

TAT (Tourism Authority Thailand) (2004a): *Internal Tourism in Phang-Nga 2006-2007.* Bangkok.

TAT (Tourism Authority Thailand) (2004b): *Internal Tourism in Phuket. 2006-2007.* Bangkok.

TAT (Tourism Authority Thailand) (2007a): *Internal Tourism in Phang-Nga 2006-2007.* Bangkok.

TAT (Tourism Authority Thailand) (2007b): *Internal Tourism in Phuket. 2006-2007.* Bangkok.

TAT (Tourism Authority Thailand) (2008a): *List Hotels Phuket Province.* Bangkok.

TAT (Tourism Authority Thailand) (2008b): *List of Hotels Phang-Nga-Province.* Bangkok.

TAT (Tourism Authority Thailand) (2009a): *TAT History.* Bangkok

TAT (Tourism Authority Thailand) (2009b): *Internal Tourism in Phuket 2007-2009.* Bangkok

TAT (Tourism Authority Thailand) (2009c): *Tourism Statistics in Thailand 1998-2007.* Bangkok

TAT (Tourism Authority Thailand) (2009d): *Internal Tourism in Phang-Nga 2007-2009.* Bangkok

TAUBENBÖCK, H.; POST, J.; ROTH, A.; ZOSSEDER, K.; STRUNZ, G.; DECH, S. (2008): *Risk and vulnerability assessment to tsunami hazard using very high resolution satellite data - The case study of Padang, Indonesia. Remote Sensing - New Challenges of High Resolution. Proceedings of the EARSeL Joint Workshop Bochum.* Bochum.

TAUBENBÖCK, H.; WURM, M.; SETIADI, N.; GEBERT, N.; ROTH, A.; STRUNZ, G.; BIRKMANN, J.; DECH, S. (2009): *Integrating remote sensing and social science– The correlation of urban morphology with socioeconomic parameters. In:* Urban Remote Sensing Joint Event, *7. Shanghai, China*

TEGTMEIER, J. (2010): *Möglichkeiten der Gebäudekartierung mit hoch aufgelösten Fernerkundungsdaten.* Diploma Thesis. München.

TELFORD, J.; COSGRAVE, J. (2007): *The international humanitarian system and the 2004 Indian Ocean earthquake and tsunamis.* In: Disasters. 31(1), 1–28.

THANAWOOD, C.; YONGCHALERMCHAI, C.; DENSRISEREEKUL, O. (2006): *Effects of the December 2004 Tsunami and Disaster Management in Southern Thailand.* In: Science of Tsunami Hazards. 24(3), 206-217.

THULER, S. (2007): *Tsunami Mitigation in Ban Nam Khem.* Worchester.

THYWISSEN, K. (2006): *Core terminology of disaster reduction. A comparative glossary.* In: BIRKMANN, J. (ed.): *Measuring vulnerability to natural hazards. Towards disaster resilient societies.* Tokyo, 448–497.

THYWISSEN, K.; WARNER, K.; RENAUD, F.; BIRKMANN, J.; VILLAGRÁN LEÓN, J. C. DE; SAKULSKI, D.; SCHLURMANN, T.; BOGARDI, J.(2006): *Disasters Reveal Vulnerability - Terms of References and Post-tsunami Activities of UNU-EHS in the Indian Ocean.* Bonn.

TICA (Thailand International Development Cooperation Agency) (2006): *Provincial Allocation of Projects.* Bangkok

TICA (Thailand International Development Cooperation Agency) (2006): *Thai Government Expenditure on Tsunami Recovery.* Bangkok.

TIERNEY, K. J. (1997): *Business Impacts of the Northridge Earthquake.* In: Journal of Contingencies and Crisis Management. 5(2), 87–97.

TIERNEY, K. J.; NIGG, J. M. (1995): Business *Vulnerability To Disaster-Related Lifeline Disruption.* DRC Preliminary Papers 223. Newark.

TIFFERT, J. (2010): *Erfassung der Tsunami people exposure an der Andamanküste Thailands. Erstellung von Datengrundlagen zur Bevölkerungs- und Touristenverteilung mit GIS & Fernerkundung.* Diploma Thesis. Kiel.

TRINH, B.; SECRETARIO, F.; KIM, K.; MANH HUNG, D. (2005): *Construction of an Inter-Regional Input-Output Table for Vietnam by the Hybrid Approach. The Case of Ho Chi Minh City and the Rest of Vietnam.* 15th International Conference on Input-Output 2005. Beijing.

TURNER, B. L.; KASPERSON, R. E.; MATSON, P. A.; MCCARTHY, J. J.; CORELL, R. W.; CHRISTENSEN, L.; ECKLEY, N.; KASPERSON, J. X.; LUERS, A.; MARTELLO, M. L.; POLSKY, C.; PULSIPHER, A.; SCHILLER, A. (2003): *A framework for vulnerability analysis in sustainability science.* In: Proceedings of the National Academy of Sciences of the United States of America. 100(14), 8074–8079.

TURVEY, R. (2007): *Vulnerability Assessment of Developing Countries: The Case of Small-island Developing States.* In: Development Policy Review. 25(2), 243–264.

TWIGG, J. (2004): *Disaster risk reduction. Mitigation and preparedness in development and emergency programming.* Humanitarian Practice Network - Overseas Development Institute. London

UNDP (United Nations Development Program) (2007): *Thailand Human Development Report 2007. Sufficiency Economy and Human Development.* Bangkok.

UNDP (United Nations Development Program) (2010): *Thailand Human Development Report 2009. Human Security, Today and Tomorrow.* Bangkok.

UNDP-BCPR (United Nations Development Program - Bureau for Crisis Prevention and Recovery) (2004): *Reducing Disaster Risk: A Challenge for Development. A Global Report.* New York.

UN-ISDR (United Nations- International Strategy for Disaster Reduction) (2004): *Living with risk: a global review of disaster reduction initiatives.* New York.

UNITED NATIONS (1991): THE ECONOMIC IMPACT OF TOURISM IN THAILAND. Economic and Social Commission for Asia and the Pacific, 1052. NEW YORK.

UNITED NATIONS; WORLD BANK (2005): *Tsunami Thailand. One year later.* New York.

UNSTATS (United Nations – Statistics Division) (2002): *National Classification – TSIC 2001.* http://unstats.un.org/unsd/cr/ctryreg/ctrydetail.asp?id=194 16.06.2011

VEEN, VAN DER, A. (2003): *In search of a common methodology on damage estimation: from the economist's perspective.* In: VAN DER VEEN, A.; VETERE ARELLANO, A.; NORDVIK J.-P. (eds.): *In search of a common methodology on damage estimation.* Twente, 4–11.

VERMAAT, J. E.; THAMPANYA, U. (2006): *Mangroves mitigate tsunami damage: A further response.* In: Estuarine, Coastal and Shelf Science. 69(1/2), 1–3.

VILLAGRÁN LEÓN, DE, J. C. DE (2006): *Vulnerability. A conceptual and methodological review.* (=Source 4/2006). Bonn.

VILLE GOYET, C. DE; MORRINIÈRE, L. (2006): *The role of needs assessment in the tsunami response.* Tsunami Evaluation Coalition. London.

VIRASA, T.; HUNT, B.; SHANNON, R.; MIN, T. Z. (2006): *Global Entrepreneurship Monitor Thailand 2006 Executive Report.* Bangkok

VOGEL, C.; O'BRIEN, K. (2004): *Vulnerability and Global Environmental Change. Rhetoric and Reality.* AVISO - Information Bulletin on Global Environmental Change and Human Security. Ottawa.

VORLAUFER, K. (2005): *Der Tsunami und seine Auswirkungen in Thailand (Teil 1).* In: Geographische Rundschau. (57)4, 14–17.

VORLAUFER, K. (2009): *Südostasien. Brunei, Indonesien, Kambodscha, Laos, Malaysia, Myanmar, Osttimor, Philippinen, Singapur, Thailand, Vietnam.* Darmstadt.

WANOUS, J. P.; HUDY, M. J. (2001): *Single-Item Reliability: A Replication and Extension.* In: Organizational Research Methods. 4(4), 361–375.

WATTS, M. J.; BOHLE, H. G. (1993): *The space of vulnerability: the causal structure of hunger and famine.* In: Progress in Human Geography. 17(1), 43–67.

WEBB, G. R.; TIERNEY, K. J.; DAHLHAMER, J. M. (2002): *Predicting long-term business recovery from disaster: a comparison of the Loma Prieta earthquake and Hurricane Andrew.* In: Global Environmental Change Part B: Environmental Hazards. 4(2/3), 45–58.

WEBBER, D.J.; ROSSOUW, S. (2010): *Sub-national vulnerability measures: A spatial perspective.* University of West England - Discussion paper 1004. Bristol.

WEERAKOON, D.; JAYASURIYA, S.; ARUNATILAKE, N.; STEELE, P. (2007): *Economic Challenges of Post-Tsunami Reconstruction in Sri Lanka.* ADBI Working Paper 75. Tokyo.

WEIBER, R.; MÜHLHAUS, D. (2010): *Strukturgleichungsmodellierung. Eine anwendungsorientierte Einführung in die Kausalanalyse mit Hilfe von AMOS, SmartPLS und SPSS.* Berlin.

WHITE, P.; PELLING, M.; SEN, K.; SEDDON, D.; RUSSEL, S.; FEW, R. (2004): *Disaster Risk Reduction. A Development Concern.* London.

WIBOONCHUTIKULA, P. (2002): *Small and Medium Enterprises in Thailand: Recent Trends.* In: Small Business Economics 18(1/3), 211–224.

WILLIAMSON, O.E. 2000: *The New Institutional Economics: Taking Stock, Looking Ahead.* In: Journal of Economic Literature. 38(3). 595–613

WISNER, B. (2002): *Who? What? Where? When? In an Emergency: Notes on Possible Indicators of Vulnerability and Resilience by Phase of the Disaster Management Cycle and Social Actor.* In: PLATE, E. J. (ed.): *Environment and Human Security: Contributions to a workshop in Bonn.* Bonn.

WOHLTMANN, H.-W. (2007): *Grundzüge der makroökonomischen Theorie. Totalanalyse geschlossener und offener Volkswirtschaften.* München.

WOLD, H. (1957): *Kausal Inferens från icke-experimentella observationer. En översikt av mål och medel.* Uppsala.

WOLD, H. (1982): *Soft modeling. The basic design and some extensions.* In: Systems under indirect observation. 2, 1–54.

WORLD BANK (2011): *How we Classify Countries.* http://data.worldbank.org/about/country-classifications 14.06.2011

WTO (World Trade Organization) (2005): *Tourism's potential as a sustainable development strategy. – WTO Tourism Policy Forum.* Madrid.

WU, S.; QIU, X.; WANG, L.; WU, S.-S.; QIU, X.; WANG, L. (2005): *Population Estimation Methods in GIS and Remote Sensing: A Review.* In: GIScience & Remote Sensing. 42(1), 80–98

WYNNE, B. (1996): *Misunderstood misunderstandings. Social identities and public uptake of science.* In: IRWIN, A. (ed.): *Misunderstanding science? The public reconstruction of science and technology.* Cambridge. 19–47

YASUDA, T. (2005): *Firm Growth, Size, Age and Behavior in Japanese Manufacturing.* In: Small Business Economics. 24(1), 1–15.

ZEIGLER, D. J.; BRUNN, S. D.; JOHNSON, J. H. (1996): *Focusing on Hurricane Andrew through the eyes of the victims.* In: Area. 28(2), 124–129.

ZHANG, Q.; FELMINGHAM, B. (2002): *An Analysis of Input-output interindustry Linkages in the PRC Economy.* Discussion paper. Hobart.

ZHANG, Y.; LINDELL, M. K.; PRATER, C. S.(2009): *Vulnerability of community businesses to environmental disasters.* In: Disasters. 33(1), 38–57.

ZOU, L.-L.; WEI, Y.-M. (2010): *Driving factors for social vulnerability to coastal hazards in Southeast Asia: results from the meta-analysis.* In: Natural Hazards. 54(3), 901–929.

ZSCHAU, J.; KÜPPERS, A. (2003): *Early warning systems for natural disaster reduction.* Berlin.

Appendix

Table 70: Outer weights for household vulnerability model (Author's design, based on own calculation)

Factor	Indicator	Weight	VIF	Durbin-Watson
Assets	Long-term consumable assets	0.30	1.10	1.93
	Property value	0.87	1.10	
Damage	Long-term consumable assets	0.52	1.11	1.74
	Property value	0.71	1.11	
Help from government	Support from government	1.00	n/a	n/a
Help from NGO	Support from NGO	1.00	n/a	n/a
Own resources	Investment of own resources	1.00	n/a	n/a
Recovery time from indirect effects	Duration of loss of income in months	1.00	n/a	n/a
Infrastructure	Interruption of road system in days	0.83	1.53	1.95
	Interruption of water system in days	0.39	1.40	
	Interruption of telephone network in days	-0.61	1.21	
Location	Distance from coastline	0.74	1.15	1.94
	Height above sea level	0.27	1.13	
	Rebuilt house	0.31	1.23	
Personal injuries	Family members life-threateningly injured	0.23	1.07	2.00
	Family members lightly injured	0.92	1.07	
Recovery time from direct effects	Time to recover long-term consumable assets	0.88	1.06	2.07
	Time to recover from property damage	0.32	1.06	
Education and income level	Average education for household members in working age	0.10	1.04	2.03
	Household income in THB	0.97	1.04	
Income from tourism	Share of household income from tourism	1.00	n/a	n/a
Income diversification	Shannon Index for income diversification	1.00	n/a	n/a

significant min 5 % error- TRAIT Household Survey

Table 71: Outer weights for company vulnerability model (Author's design, based on own calculation)

Factor	Indicator	Weight	VIF	Durbin-Watson
Assets	Amount of productive assets and property	1.00	n/a	n/a
Location	Distance from coastline	0.35	1.12	2.23
	Height above sea level	0.83	1.12	
Damage	Total damage to productive assets and property	1.00	n/a	n/a
Credit	Total credit volume 2007	1.00	n/a	n/a
Physical recovery	Time to recover from property damage	0.56	1.32	2.15
	Time to recover productive assets	0.63	1.32	
Education level	Education of management in years	1.00	n/a	n/a
Investor	Percentage of domestic investment	0.66	1.04	1.40
	Percentage of foreign investment	0.81	1.04	
Ext. help	External support for property value	0.79	1.00	2.01
	External support for productive assets	0.61	1.00	
Age	Company age	1.00	n/a	n/a
Indirect effects	Interruption of infrastructure in days	0.97	1.00	2.15
	Absence of workforce in days	0.16	1.00	
Supply	Supply affecting business activities	0.69	1.11	1.52
	Days of interruption of supply	0.54	1.11	
Location of supplier	Major supplier located in Phang-Nga	1.00	n/a	n/a
Location of sales	Major market in Phang-Nga	1.00	n/a	n/a
Backward linkages	Backward linkages in percentage	1.00	n/a	n/a
Recovery of sales	Days of reduced sales	1.00	n/a	n/a
Market environment	Company subject to seasonal changes	1.00	n/a	n/a
Dependency on tourists	Percentage of sales to tourists	1.00	n/a	n/a
Integration in value chain	Forward linkages in percent	0.55	1.18	1.96
	Backward linkages in percentage	0.98	1.18	

significant min 5% error - TRAIT Company Survey

A - List of focus groups

Abbreviation	Explanation
FG-Pla	Focus group plantations
FG-Fish	Focus group fishing
FG-Inc	Focus group incidental farming
FG-Hot	Focus group hotels
FG-Res	Focus group restaurants
FG-Tou	Focus group tour agencies
FG-Ret	Focus group tourism-related retail

Abbreviations are extended with the following study area codes:
Ban Nam Khem – BNK; Khao Lak – KL; PB – Patong Beach;

B - Interview sources

KHORPETCH, P.; Vice President of the Hotel Association Khao Lak, Guided interview on 23.09.2009 by Revilla Diez, J.

NANTANA, Owner of Phuket Merlin Hotel, Guided interview on 25.09.2009 by Revilla Diez, J.

RUNAGAPIBAL, K. Owner of Khao Lak Ecotourism, Guided interview on 24.09.2009 by Willroth, P.